HUMAN BIOLOGY

HUMAN BIOLOGY

Elementary Anatomy and Physiology
for Students and Nurses

JOHN GIBSON, M.D.

Third Edition

FABER AND FABER
LONDON   BOSTON

First published in 1960
by Faber and Faber Limited
3 Queen Square London WC1
Second edition 1967
First published in this edition 1972
Reprinted with amendments 1975
Third edition 1978
Phototypeset in VIP Times by
Western Printing Services Ltd, Bristol
Printed in Great Britain by
Whitstable Litho Ltd Whitstable Kent

British Library Cataloguing in Publication Data

Gibson, John, b.1907
   Human biology. – 3rd ed.
   1. Physiology
   I. Title
   612   QP36

ISBN 0–571–04974–5 (papercovers)

CONTENTS

1.	General Principles and Basic Tissues	*page* 13
2.	Growth before Birth	20
3.	Growth after Birth	25
4.	Bones, Joints and Muscles	31
5.	Head, Neck and Trunk	38
6.	Arms and Legs	57
7.	The Circulation of the Blood	73
8.	Respiration	85
9.	The Blood	93
10.	The Alimentary Tract	100
11.	Nutrition and Metabolism	113
12.	The Liver and the Pancreas	121
13.	The Urinary System: Water Balance	127
14.	The Skin: The Temperature of the Body	136
15.	The Lymphatic System	141
16.	The Nervous System	145
17.	The Special Senses	168
18.	The Endocrine Glands	177
19.	Reproduction and Heredity	186
20.	Defence Mechanisms	199

PREFACE

For the third edition this book has been revised and brought up to date, and many sections have been largely rewritten.

FIGURES

		page
1/1	Epithelial cells	15
1/2	Muscle cells	17
2/1	An ovum and spermatozoon	20
2/2	The development of the fetus	23
3/1	The epiphyses and centres of ossification in the radius	27
3/2	An infant's skull at birth	28
3/3	The first and second dentitions in a child of six years	29
4/1	The bone structure of the upper end of the femur	31
5/1	The skull – lateral view	38
5/2	The inside of the left half of the skull	39
5/3	Structures in the face and neck	42
5/4	The vertebral column	43
5/5	Thoracic vertebrae	44
5/6	The chest and diaphragm	47
5/7	The skeleton – anterior view	50
5/8	The pelvis	51
5/9	The muscles of the body – anterior view	53
5/10	The skeleton – posterior view	54
5/11	The muscles of the body – posterior view	56
6/1	The shoulder girdle	57
6/2	The elbow-joint	61
6/3	The arteries of the arm	63
6/4	The hip-joint	66
6/5	The knee-joint	67
6/6	Structures forming the arch of the foot	69
6/7	The arteries of the leg	71
7/1	A diagram of the circulation	74
7/2	A diagram showing the heart and main vessels	75
7/3	A normal electrocardiogram	79
7/4	Valves in a vein	83
8/1	A section through the respiratory passages	86
8/2	The heart and lungs	88

8/3	Blood circulation over the alveoli of the lungs	89
8/4	The chest wall in inspiration and expiration	91
9/1	Red and white blood-cells and platelets	95
10/1	The mechanism of swallowing	101
10/2	The digestive tract	103
10/3	The stomach, empty and full	104
10/4	The structures at the back of the abdomen	106
10/5	The peritoneum, in the female body	111
12/1	The liver cells and sinusoids	122
12/2	The biliary system, pancreas and pancreatic duct	123
13/1	The urinary system	128
13/2	A nephron	129
14/1	A section through the skin	137
15/1	A lymph gland and vessels	142
15/2	The thoracic duct	143
16/1	A nerve cell, showing a synapse	146
16/2	Lateral view of brain	148
16/3	A section through the brain	152
16/4	A section through the spinal cord	155
16/5	The position of the ventricles within the brain	157
16/6	The skull and meninges and a venous sinus	158
16/7	The blood supply to the brain	160
16/8	An electro-encephalogram	162
17/1	Lateral view of structure of eye	169
17/2	The lacrimal system	171
17/3	The optic nerves	172
17/4	The structure of the ear	173
18/1	The pituitary gland and pituitary fossa	178
18/2	A diagram of the functions of the pituitary gland	179
18/3	The thyroid and parathyroid glands	181
19/1	The male genital organs	187
19/2	A section through the female pelvis	189
19/3	The female genital organs	191
19/4	A diagram of the menstrual cycle	193
19/5	The structure of the breast	195

Chapter 1

GENERAL PRINCIPLES AND BASIC TISSUES

Anatomy is the study of the structure of the body. **Histology** is the study of the microscopic anatomy of the body. **Embryology** is the study of the development of the body before birth. **Physiology** is the study of the functions of the body. **Biochemistry** is the study of the chemistry of living tissues.

The study of form which is anatomy cannot be separated from the study of function which is physiology, for the structure and function of any part of the body are closely related. We can see this in a study of the arm and leg.

The *arm* is an organ for feeling and grasping. The hand must be light, flexible, well endowed with sensory nerve-endings, particularly in the finger-tips, and able to grasp tightly or gently; the bones must be light, and the whole structure must be capable of a wide range of movements, which is largely achieved by the mobility of the shoulder joint.

The *legs* have a different function. They have to support the weight of the trunk, head and arms, and they have to move the body from place to place. In consequence the bones in them are thick and strong, the joints tight, and the muscles big and powerful.

THE CELL

The *cell* is the structural unit of which all tissues are composed. All the cells of the body are derived from the union of an ovum and a sperm, and each cell develops according to its function into its appropriate size, shape, physical appearance and chemical composition.

Each cell is basically composed of:

1. *A cell membrane*, which is a thin membrane enclosing the rest of the cell.
2. *Cytoplasm*, which is a tiny blob of jelly-like material.
3. *A nucleus*, which is a small dense structure lying within the cytoplasm.
4. *A nucleolus*, which is a small body lying within the nucleus.

The basic features of the functioning of any cell are:

1. All cell activity requires oxygen and produces carbon dioxide.
2. The cell membrane has selective permeability, i.e. it allows certain substances (but not all) to pass through it in either direction, into or out of the cell.
3. Enzymes (which speed up chemical reactions without themselves being changed in the process) are present in the cytoplasm, and each of them promotes a specific chemical reaction within the cell.
4. DNA (deoxyribonucleic acid) and RNA (ribonucleic acid) are present in cells and are necessary for their reproduction.

Most of the cells of the body can reproduce themselves by splitting into two, the nucleus splitting first and then the rest of the cell, and the two cells thus formed grow rapidly to their full size. Most cells are fixed in one position, but some cells move about – blood cells are moved in the bloodstream around the body, and some cells called histiocytes are capable of amoeboid-like movements.

THE TISSUES OF THE BODY

The cells of the body form five basic kinds of tissue:
1. Epithelium
2. Connective tissue
3. Skeletal tissue
4. Muscle
5. Nervous tissue

The cells of these tissues appear in them in many different forms.

1. Epithelium

Epithelial cells form the skin, the inner lining of the heart, blood-vessels and lymphatics, the inner lining of the digestive system, the alveoli (air sacs) in the lungs, the urinary system and the genital system, and the secreting cells and ducts of glands (see Fig. 1/1)

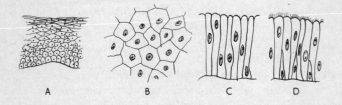

A   B   C   D

Fig. 1/1 Epithelial cells: (a) stratified epithelium; (b) pavement epithelium; (c) columnar epithelium and (d) ciliated epithelium

Stratified epithelium in which the cells are arranged in several layers, is present in the epidermis, the outer layer of the skin.

Pavement epithelium which forms the smooth inner lining of the heart, blood-vessels and lymphatics.

Ciliated epithelium which is present in the respiratory tract and the uterine tubes. The cells are columnar and *cilia*, thin hair-like processes project from their free surface and by moving to and fro cause rippling movements to pass over the surface. In the respiratory passages they cause any dust that lands on them to be driven away from the lungs; and in the ovarian tubes they propel the ovum towards and into the uterus.

2. Connective Tissue

Connective tissue supports and connects tissues more active than itself.

Areolar tissue which is composed of interlaced elastic threads, forms the sheaths that enclose muscles, nerves, blood-vessels and other structures, and allows one structure to move easily over another.

Adipose tissue which is composed of rounded fat cells, held in a network of areolar tissue.

Fibrous tissue which forms the tendons of muscles, the ligaments that bind bones at joints, and the capsules of joints.

3. Skeletal Tissue

Skeletal tissue forms the bones and cartilage.

Bones are made of tough, hard tissue, enclosed within the periosteum, a layer of fibrous tissue. *Cartilage* occurs in three forms: *hyaline cartilage* which forms the costal cartilages at the front ends of the ribs and covers bone where it is enclosed in a joint; *white fibrocartilage*, which forms the intervertebral discs of the spinal column and cartilages in the knee and other joints; and *yellow fibrocartilage*, which forms the cartilaginous parts of the nose and ears.

4. Muscle

Muscle cells have the ability to contract and relax. They are commonly called muscle fibres. There are three kinds of muscle: voluntary, involuntary and cardiac (see Fig. 1/2).

Voluntary (striped) muscle is formed of slim, transversely striped fibres, varying in length from one to forty millimetres; they are attached to bone or cartilage, and are under conscious control.

Involuntary (plain or unstriped) muscle occurs in the stomach, intestines, bladder, the blood-vessels and other internal organs. The fibres are short and unstriped. Their contraction and relaxation are controlled through the autonomic nervous system.

Cardiac muscle is present only in the heart. The fibres are short, striped, and connected to one another by strips of muscle, so that the whole forms a continuous structure. Their contractions and relaxations are automatic and involuntary.

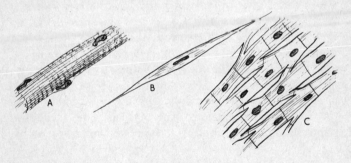

Fig. 1/2 Muscle cells: (a) striped muscle; (b) plain muscle and (c) cardiac muscle

5. Nervous Tissue

Nervous tissue forms the brain, spinal cord, the nerves attached to them, and the autonomic nervous system. The cells in it are:

1. *Neurones*   Nerve cells and the fibres attached to them; they are the essential cells which carry out nervous functions.
2. *Neuroglia*   Cells whose exact function is not known but may be to absorb and destroy micro-organisms and foreign substances which have got into nervous tissue.

THE LIFE OF A CELL

A living cell is in a state of constant activity and its chemical composition is being built up and broken down all the time. The activity and life of a cell depend upon its working conditions being kept stable. It has to be provided with sufficient oxygen for its needs, and if it does not get enough it starts to function badly; and if the shortage of oxygen is prolonged it dies. The chemical composition of the fluid bathing a cell has to be kept within a very narrow range. The temperature to which it is exposed must not vary more than a few degrees, whatever activity is going on in the body or whatever the temperature of the air around the body. It must be provided with adequate amounts of water, food, minerals and vitamins, according to its precise needs. The waste products, such as carbon dioxide, produced by it in the course of its chemical activities, must be promptly removed and not allowed to accumulate within or around it.

As a result of the total activities of his cells a person is able to live. To live he must be able to react to his environment. He must be aware through his senses of what is going on around him and particularly of any danger that threatens him. He must be able to avoid danger by making appropriate muscular movements. He must be able to absorb foods, to breathe, and to excrete waste-stuffs. He must be able to control the internal economy of his body, keeping its chemical composition and temperature within narrow limits. The health of his tissues must be properly controlled by secretions from his endocrine glands. He must be able to co-ordinate his activities by the use of his nervous system. He will have reproductive powers and the urge to reproduce.

Chapter 2

GROWTH BEFORE BIRTH

Fertilisation

A human being is created by the fertilisation of an ovum, the female sex cell, by a spermatozoon (sperm), the male sex cell (see Fig. 2/1). The event takes place in one or other of the uterine tubes, which open at one end over an ovary (in which the ova are formed) and at the other into the uterus. Immediately, in the fertilised ovum, a number of rapid changes take place. The single cell produced by the fertilisation promptly splits into two; each of these splits into two more, the four cells into eight, and so on. This reproduction of cells goes on to form a little, tightly packed mass of cells, out of which are to come a human being and

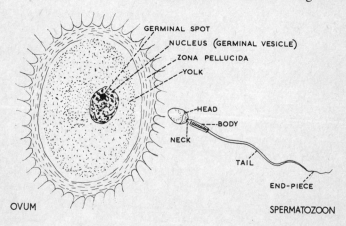

Fig. 2/1 An ovum and spermatozoon

the membranes and attachments that are to serve and protect him until he is born.

Twins occur about once in eighty births. Identical twins, of the same sex and appearance, are formed if, as happens occasionally, the first two cells into which a fertilised ovum divides happen to separate one from the other, so that two beings are formed. Dissimilar twins, who do not look alike and need not be of the same sex, are due to two ova being fertilised at the same time by two sperms.

Development of the Embryo

The mass of cells that is to be a person is moved by ciliary action along the uterine tube, and about the third or fourth day after fertilisation is discharged out of the tube into the uterus, where it is to live until birth. The endometrium (the inner lining of the uterus) has thickened to receive it, and into its soft, blood-engorged tissue the mass of cells sinks and sticks. The *embryo* that is to be a child has by now appeared as a distinct oval mass and becomes enclosed by membranes in a pool of fluid, the amniotic fluid. From the membranes finger-like processes grow into the endometrium. The *placenta* is formed from these processes, a soft, flat disc of blood-vessels, attached on one side to the endometrium and on the other connected to the embryo by a stalk, which is to become the umbilical cord. Through the placenta and cord the growing child can draw from the blood of his mother the oxygen and food he requires and can discharge the carbon dioxide and other waste products that he produces (see Fig. 2/2).

At the third week of existence the embryo is recognisably a little creature, with a head at one end and a bulge over the heart. Three types of tissue are now distinguishable:

the *ectoderm* (on the outside) which is to form the skin and nervous tissue;

the *mesoderm* (in the middle) which is to form the bones,

muscles, heart, blood-vessels, blood, kidneys and sex glands.

the *endoderm* (on the inside) which is to form the alimentary tract, the lungs, the liver and the pancreas

The **nervous system** is formed out of a long furrow which appears along the back of the embryo, the edges of the furrow eventually fusing to form a tube, the neural tube, which is to develop into the brain, the spinal cord and their nerves. The fore-brain, the mid-brain and the hind-brain appear as three little knobs at the head-end of the tube, the fore-brain growing much more rapidly than the others and growing over the top of them.

The **heart** forms behind the enlarging head by the union of several arteries and starts beating spontaneously when the embryo is barely four weeks old. The changes that the heart has to make in its construction are doubly complicated, for it has to adapt itself to two modes of existence: the first in the unborn child whose lungs are not working and whose blood has to go to and from the placenta; the second immediately after birth when the blood has to be pumped through the now active lungs and the holes in the heart through which it had previously passed have to close.

The **arms** and **legs** appear as little buds about the fourth week, the buds for the arms appearing a little before those for the legs. In time the fingers and toes appear, and bone, cartilage and muscle develop in the limbs, and nerves grow down into them.

The **alimentary tract** develops as a tube from the mouth, just between the developing brain and heart, to the anus between the thighs. At first both ends are closed off. Within the abdomen the tube enlarges into the stomach and coils itself into the small and large intestines. The **lungs** are an outgrowth from the upper end of the tract, and as they grow downwards they push in front of them a strip of muscle that is to form the diaphragm. The **liver** and **pancreas** develop within the abdomen as outgrowths from the alimentary tract.

The kidneys, the bladder, the testes (male sex glands) and the ovaries (female sex glands) are formed from masses of mesodermal tissue at the back of the abdomen. The ovaries migrate downwards into the pelvis. The testes have

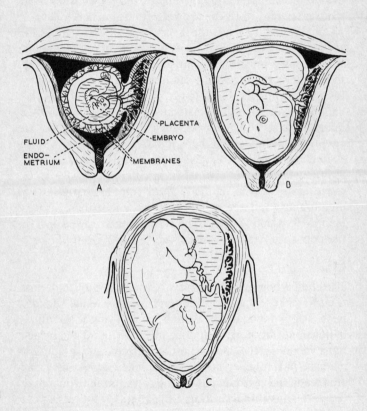

Fig. 2/2 The development of the fetus: (a) an embryo at the sixth week; (b) an embryo at the seventh week, and (c) a fetus at the fifth month

a longer course to take: they descend down the back of the abdomen and then out of it to appear in the scrotum about the time of birth.

Development of the Fetus

The major changes in the construction of the various organs are completed by the end of the eighth week, and from now until birth the growing child is called a *fetus*.

At the eighth week the fetus is a little person, with appropriate organs on a minute scale. It is only about 30mm long. From now onwards the fetus grows rapidly and makes heavy demands on his mother for the food he needs: for oxygen, for the basic foods, for minerals such as calcium, phosphorus and iron, and for vitamins. If the mother does not take adequate amounts of these in her diet to supply the needs both of herself and her child, they will be drained off her own tissues for the child's use.

When the fetus is about five months old, he begins to move about in the amniotic fluid contained within the membranes, and the mother starts to feel these movements. He usually comes to lie with his head downwards in the pelvis waiting to be born (see Fig. 2/2(c)).

Birth

The child is born about 266 days after conception, a period usually reckoned as nine calendar months. What sets off birth is not known; it is likely to be changes in the secretion of hormones from the endocrine glands. Muscular contractions begin in the muscle of the uterus and gradually increase in frequency and strength. The lower end of the uterus and the vagina expand. Under the pressure put upon them, the membranes rupture, the amniotic fluid escapes, and after a labour which usually lasts for several hours, the baby is born. The placenta and the membranes are discharged, as the 'after-birth', about fifteen minutes later.

Chapter 3

GROWTH AFTER BIRTH

At his birth the baby has to adjust rapidly to new circumstances. For the first nine months of his life he has been carefully protected, cushioned by fluid and muscle, fed without any effort on his part, provided automatically with all he has required, kept in darkness, and protected from most of the harmful micro-organisms; and now he is exposed to the difficulties and dangers of independent life.

For a few minutes after birth the child is still attached by the umbilical cord and placenta to the interior of his mother's uterus. The doctor or midwife ties the umbilical cord and divides it with scissors, and the placenta and membranes are expelled from the uterus. To lead an independent life the baby must breathe. He gasps, fills his lungs with air, and continues with the cycle of breathing. Changes have to take place in his heart. Two temporary passages (the foramen ovale between the right and left atria, and the ductus arteriosus from the pulmonary artery to the aorta) have to close, a process that takes several days, in order that the blood can now be directed through the lungs. If these changes do not take place, the baby will have a congenital disease of the heart, and will be a 'blue baby' if his blood is prevented from becoming adequately oxygenated in the lungs.

The umbilical cord shrivels up, the stump remaining as the umbilicus.

For some time after birth the chemical composition of the baby's blood resembles that of his mother, containing, for example, adult hormones. After a few days or weeks these are broken down, and the blood chemically becomes that of a child.

At birth a baby's weight is usually 3 to 4kg (6–9lb). He is likely to be underweight if his mother is small or is a heavy smoker or drinker, if the birth has been premature or he is one of twins or triplets. He loses a few grams of weight during the first few days of life and should regain his birth-weight in ten days. He should double his birth-weight in six months and treble it in twelve. At birth his body-length is about 50cm; it should be doubled in twelve months.

The Brain

The brain grows rapidly during the first year, and then more slowly until six years when it is fully grown. This growth is not due to an increase in the number of nerve-cells, which are all there at birth, but to myelination of the nerve-fibres, to an increase in the number of neuroglia, and to an increase in the size of the blood-vessels.

Myelination is the formation of a sheath of fatty material around the long fibre of a nerve-cell. It begins during fetal life and is completed after birth. It is thought that a nerve fibre cannot function correctly until it has been myelinated. The fibres conveying the sensations of touch, sight and sound are the first to be myelinated, and then the motor fibres.

Bones

The bones grow in length and thickness. Most of them grow out of sticks of cartilage. The centre of ossification is the place in a piece of cartilage where bone formation begins. One of these centres of ossification appears in the middle of the cartilage and forms most of the bone. At birth the shafts of the long bones are already bone. Cartilage is then still present at the ends of these bones, and one or more centres of ossification appear in them. An *epiphysis* is one of these separate pieces of bone; at birth the only epiphysis that has appeared is that at the lower end of the femur in the thigh. An epiphysis is separated from the bone of the shaft by a piece of cartilage, the presence of which can be seen on an X-ray as an *epiphyseal line*. Both bones grow into this piece of cartilage, and so long as the cartilage remains the bone can increase in length. Eventually the two pieces of bone fuse through the cartilage, and when this has happened the bone cannot grow any longer. As the dates at which epiphyses normally appear and fuse with the shaft are regular, it is possible to tell the age of a young skeleton by X-ray examination. The radius, one of the bones in the forearm, is, for example, made in this way (see Fig. 3/1):

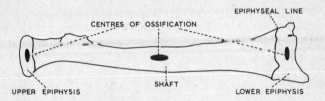

Fig. 3/1 The epiphyses and centres of ossification in the radius

1. A centre of ossification appears in the shaft at the eighth week of fetal life.
2. A centre of ossification appears for the upper epiphysis at four years and fuses with the shaft at seventeen years.
3. A centre of ossification appears for the lower epiphysis at one year of life and fuses with the shaft at twenty years.

The radius is therefore not fully completed until the twentieth year.

A bone grows in thickness by the formation of new bone from cells in the inner surface of the periosteum which surrounds it.

The *legs* and *pelvis* are small at birth, and do not develop very much until the baby starts to stand up and to walk. The foot is flat at first, and the arches of the feet develop with standing and walking.

Skull

The bones of the top of the skull develop from membrane, not cartilage, and at birth are still separate thin sheets of bone. The *anterior fontanelle* is a lozenge-shaped piece of membrane at the spot where the two frontal bones and the two parietal bones are going to meet (see Fig. 3/2); it yields a little when pressed with a finger and does not ossify until a baby is about eighteen months old. The *posterior fontanelle* is a similar but smaller area at the back of the head where the occipital bone meets the parietal bones; it ossifies shortly after birth. The fontanelles, the thinness of the bones, and the presence of membrane between individual bones allow 'moulding' (a slight change of shape) of the skull to take place during the birth of the head.

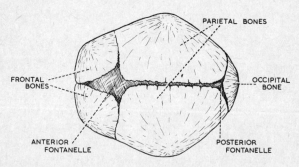

Fig. 3/2 An infant's skull at birth seen from above

The *teeth* grow from tooth-buds in the jaws. The first dentition consists of twenty 'milk' teeth. The first appears at six months and the twentieth at about two years. At six years the crowns of these teeth start to break off and the roots to be absorbed (see Fig. 3/3). They are replaced by the thirty-two permanent teeth of the second dentition, which are larger and stronger. The first of these appears at six years and the last – the 'wisdom teeth' – between twenty-one and twenty-five years. If the wisdom teeth do not appear, they are called unerupted.

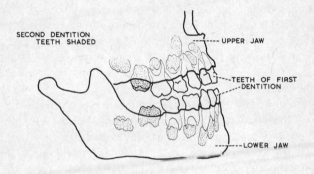

Fig. 3/3 The first and second dentitions in a child of six years

Puberty

Puberty is the time of the beginning of sexual maturity. A number of skeletal and other changes take place at the same time. It occurs in girls at about ten to twelve years and in boys about twelve to fourteen years. At this stage of development girls are taller than boys of the same age.

In girls the ovaries are stimulated into activity by the pituitary gland. The genital organs enlarge; menstruation begins; the breasts develop; pubic and axillary hair starts to grow; fat is laid down; the pelvis becomes broader.

In boys the testes are stimulated into activity by the pituitary gland. The genital organs enlarge; sperm are formed; facial, pubic and axillary hair starts to grow; the larynx doubles in size and the voice 'breaks' and becomes deeper; a spurt in growth starts.

Adolescence

Adolescence is the period between puberty and reaching adult life.

During this period a girl grows about 4cm and a boy about 10cm, and boys now become taller than girls of the same age. Full height is reached at about twenty years when the epiphysis at the lower end of the femur fuses with the shaft of the bone. But bones can continue to grow in thickness from cells on the inner layer of the periosteum.

Adult Life

At the age of twenty years men and women are fully grown. They have now reached their maximum physical strength, which they may retain for forty or fifty years.

Various changes take place in the skeletal tissues. The thin sutures which have separated the bones of the top of the skull ossify from the age of thirty onwards, and the parts of the sternum (breast bone) over years become ossified into one bone.

The *menopause* occurs in women between forty and fifty years. The cells of the ovary cease to respond to stimulation by the pituitary gland; menstruation stops and the genital organs become smaller.

Old Age

In old age people can become smaller as a result of a shrinking of the intervertebral discs in the spine. Weight is reduced. Calcium is lost from the bones, which become more liable to fracture. With loss of teeth the mandible, (lower jaw bone), becomes slim. The skin loses some of its elasticity. The ability to hear high-pitched sounds diminishes. Intelligence can decline sharply.

Chapter 4

BONES, JOINTS AND
MUSCLES

BONES

The skeleton is a jointed structure of bones and cartilages.
Bones are composed of:

(a) *Compact bone* which is a dense, hard layer of bone on the
   outside, the bone being arranged along long canals
   called Haversian canals;

(b) *Cancellous bone* which is a honeycomb of thin struts of
   bone inside the compact bone (see Fig. 4/1);

(c) A *medullary cavity* which runs down the middle of long
   bones, and is filled with bone marrow;

(d) *Periosteum* which is a tough membrane firmly attached
   to the exterior of bone, except within joints; bone cells

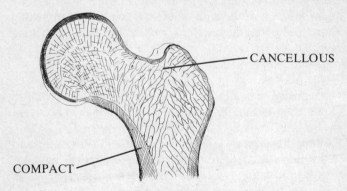

Fig. 4/1 The bone structure of the upper end of the femur

grow from the innermost layer of the periosteum, increasing the thickness of a bone during the growing period and repairing a fracture if it occurs.

When a dried bone is examined, the compact bone, the cancellous bone and the medullary cavity can be seen; but if a fresh, young bone with a joint is obtained from a butcher, the toughness and tight attachment of the periosteum can be appreciated, the smoothness and moisture of the pearly-grey cartilage in the joint felt, and bone marrow seen in the cancellous bone and medullary cavity.

Bones are composed partly of inorganic salts, especially calcium phosphate, which makes them strong as well as opaque to X-rays, and partly of organic material, especially proteins, the fibrous tissue in them giving toughness and resilience.

Bones are living structures. In childhood they have to grow into adult size; throughout life they are in a state of being constantly broken down and rebuilt; if they are broken they repair themselves; they have important functions. They need a good supply of blood. Blood vessels enter and leave a bone at the nutrient canal, a hole about half way along the shaft of a long bone and through several holes near the ends of the bone where the epiphyses have had to have been well supplied with blood during the growing period.

Bone Marrow

Bone marrow is a mushy substance present in cancellous bone and the medullary cavities of long bones. It occurs in two forms:

(a) *Red bone marrow* in which red blood cells and some white blood cells are formed; it is present in all bones in the fetus and young child when there is a great need for cells for the increasing amount of blood, but after about the fourth year it is progressively replaced by yellow bone marrow, and by adolescence it is present only at the upper ends of the humerus and femur, the sternum, the ribs, the bodies of the vertebrae, and the flat bones of the skull.

(b) *Yellow bone marrow* which is composed of unimportant fatty and connective tissue which has replaced red bone marrow.

Functions of the Skeleton
The functions of the skeleton are:
1. To form the framework of the body.
2. To provide levers for muscles to act on.
3. To protect organs: the skull protects the brain, the ribs protect the heart and lungs, the pelvis protects the bladder and rectum and in women the internal genital organs.
4. To form red blood cells and some white blood cells.
5. To be a store of calcium and phosphorus, which are deposited in bone from the blood and withdrawn from it when required in the rest of the body.

JOINTS

A joint is present where two or more bones meet. There are three main types of joint: the movable joints, the slightly movable joints and the immovable joints.

Movable Joints
Movable joints occur mainly in the limbs, which are the parts of the body in which maximum movement is necessary. Although the joints differ very much in details, according to their functions and the amount of movement needed, they all have a certain basic structure:
1. The bones forming the joint are separate and their ends within the joint are covered with hyaline cartilage.

2. The joint is totally enclosed within a capsule of fibrous tissue.
3. A smooth synovial membrane lines the inside of the capsule and secretes synovial fluid which keeps the inside of the joint moist, thus allowing one surface to glide smoothly over another.
4. Ligaments, composed of strong fibrous tissue, are present on the outside of the capsule, strengthening the joint and preventing too much movement being made.

Some joints, such as the knee joint, have discs of fibrocartilage between the bones; such cartilages are likely to be torn if too much stress is put on the joint. Bursae are present in some joints; a bursa is a little completely enclosed bag, which assists in diminishing friction at a spot in the joint.

Some joints, such as the elbow, move like a hinge; some, like the shoulder and hip joints, move like a ball in a cup; some, like the joints between the little bones in the wrist, have little gliding movements. How much a joint can move is controlled by a number of factors – the shape of the bones at the joint, the attachment of ligaments, the position and strength of muscles acting on the joint, and the meeting of soft parts, such as the thigh on the anterior abdominal wall. The stability of a joint is due in large measure to the presence and strength of the muscles and tendons that act on it, but movable joints are in different degrees unstable and liable to be dislocated when subjected to excessive strain.

Slightly Movable Joints

Slightly movable joints are present between adjacent vertebrae and at the pubic joint in the pelvis, where a little movement takes place over a central pad of cartilage. This pubic joint widens a little during pregnancy and childbirth in order to enlarge the pelvis and let the baby through with less difficulty.

Immovable Joints
Typical immovable joints occur at the sutures of the skull, where the bones meet and are firmly interlocked at irregular saw-like edges. These sutures ossify in old age.

MUSCLES

Skeletal muscle (which is also called striped or voluntary muscle) forms about 40 per cent of the total weight of the body. It is employed in holding the body erect and in making voluntary movements.

A muscle is composed of muscle fibres, thin cells varying in length from 1 to 40mm. Each fibre is composed of smaller fibres called myofibrils. Fibres and myofibrils, seen through a microscope, show striations, that is a number of alternate light and dark bands. Some of these bands become narrow when a muscle contracts.

To contract a muscle must be attached at both ends. Most of them are attached to bone, either directly into the periosteum or by means of fibrous tissue. Some muscles end by becoming a tendon, a strong cord of fibrous tissue which is inserted at its other end into bone. A tendon enables a large muscle to concentrate its strength on a single small area of bone; several tendons can be present in a small space (as at the wrist) where there would not be room for a muscle; and tendons provide protective and supportive action around the joint. Where a muscle is attached directly into periosteum, the surface of the bone is smooth, but where it is attached by a mixture of muscle and fibrous tissue or by fibrous tissue or by a tendon, the bone is likely to be rough and show ridges and knobs.

Contraction of Muscle

A muscle contracts by the shortening of the individual fibres of which it is composed, and it contracts in response to stimuli reaching it through a nerve.

A *neuromotor unit* is a unit composed of a motor nerve fibre and the muscle fibres it supplies. A motor nerve is one down which pass stimuli from the brain or spinal cord. A motor nerve fibre divides into a number of smaller fibres, each of which supplies an individual muscle fibre, ending on it in a motor end-plate, a flat structure lying against the muscle fibre. When a nervous impulse reaches the motor end-plate, acetyl choline, a chemical substance, is released from it and causes the fibre to contract. After contracting the fibre relaxes and returns to its resting length. In the contraction of a fibre, oxygen is used up, carbon dioxide is formed, and heat is produced.

No single muscle can act by itself. Muscles act in groups, and by the system called *reciprocal innervation* when one group of muscles contract, the muscles with the opposite action relax to the same extent, the degree of movement being regulated by these two forces. Many other muscle groups are likely to be indirectly involved and have to make some adjustment by contraction or relaxation. If the arm is bent at the elbow, the biceps muscle in front contracts, the triceps behind relaxes, and muscles around the shoulder joint have to make adjustments to keep the shoulder joint steady.

Muscle Tone

Muscle tone is the slight degree of tension present in muscle all the time. It is necessary to act against gravity and to maintain an upright position. It is produced by nervous impulses which pass down the motor nerves to a small number of motor units at any one time, the units in action being repeatedly changed so that fatigue is prevented. A muscle is not shortened by the tone in it. Sensory nerve endings in muscles and tendons relay through sensory nerves to the brain and spinal cord the degree of tension in a muscle so that they may respond by sending down an appropriate number of motor impulses.

Chapter 5

HEAD, NECK AND TRUNK

THE HEAD AND NECK

The skull consists of two parts – the cranium and the mandible (lower jaw) (see Fig. 5/1).

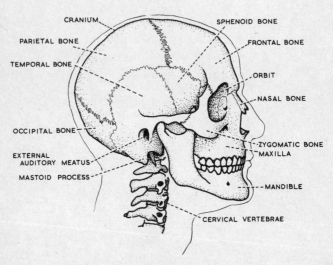

Fig. 5/1 Lateral view of the skull

THE CRANIUM

The cranium consists of several bones fixed together by immovable suture joints. The cranial vault roofs over the brain; the base of the skull is the part on which the brain rests; the bones of the face form the rest of the cranium.

The *cranial vault* (skull cap) is thin and formed of: (a) the frontal bone in front; (b) the right and left parietal bones at the top and sides; (c) the occipital bone at the back, and (d) parts of the sphenoid and temporal bones at the sides.

The *base of the skull* is thick and formed of: (a) the frontal bone in front; (b) the ethmoid bone; (c) the sphenoid bone; (d) the right and left temporal bones, and (e) the occipital bone at the back.

It is pierced by many holes for nerves, arteries and veins, and by the foramen magnum, a large hole in the occipital bone for the spinal cord (see Fig. 5/2). On the under surface of the skull, on either side of the foramen magnum, are the joint surfaces for the first cervical vertebra (the atlas).

The *front of the skull* shows: (a) the frontal bone, which forms the forehead; the orbits, (the sockets for the eyes); (b) right and left nasal bones, which form the bridge of the

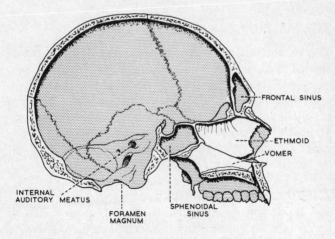

Fig. 5/2 The inside of the left half of the skull

nose; (c) the nasal openings, divided by a central septum; (d) the right and left maxillae (upper jaw bones), which carry the upper teeth, and (e) the right and left zygomatic bones, which connect maxilla and temporal bone, forming the prominence of the cheek.

At the *side of the skull* the temporal bone shows: (a) the external auditory meatus, a bony tube leading to the middle ear, and (b) the mastoid process, a rounded prominence of variable size, palpable just behind the ear.

The Accessory Air Sinuses

The accessory air sinuses are cavities of varying size and shape in the maxillae, frontal bones, ethmoid bone and sphenoid bone. The air sinus in a maxilla is called a maxillary antrum. They act as resonators to the voice and lighten the weight of the skull. They all have openings into the nasal cavities, the mucous membrane which lines them is continuous with that lining the inside of the nose, and infections of the nose can readily pass into a sinus.

THE MANDIBLE

The mandible (lower jaw bone) is a strong horseshoe-shaped bone which carries the teeth of the lower jaw. Behind the teeth it bends upwards to end in two processes, a pointed one in front for a muscular attachment, and a rounded one behind, articulating with the temporal bone at the temporo-mandibular joint. A violent yawn will sometimes dislocate the mandible forwards out of the slight depression in the temporal bone.

The Hyoid Bone

The hyoid bone is a delicate, slender, horseshoe-shaped bone in the neck, just above the thyroid cartilage (Adam's apple). It is not attached to any other bone, being held in position by muscular and fibrous attachments to the tongue above and the larynx below (see Fig. 5/3).

The Teeth

The *first dentition* is composed of twenty teeth: four incisors, two canines and four molars in each jaw. The *second dentition* is composed of thirty-two teeth: four incisors, two canines, four premolars and six molars in each jaw.

The incisors have a cutting edge; the canines are longer and slightly pointed; the premolars have two rounded projections called cusps and two roots; the molars are much larger teeth with two or three cusps and two or three roots. Each tooth is composed of:

enamel, the white insensitive surface of the tooth within the mouth;

dentine, which is very sensitive, inside the enamel;

a pulp of blood-vessels and nerves in a central cavity;

cement, which attaches the tooth to the jaw.

The Muscles of the Head and Neck

The *muscles of facial expression* are a number of small, flat or slim muscles in the face, scalp and neck. They are used to express emotion and shut the mouth and eyes (see Fig. 5/3).

The *muscles of mastication* are stronger and larger muscles which move the mandible on the cranium and are used in opening and closing the mouth and in biting and chewing.

The *muscles of the neck* are used for supporting the head and spine, for moving the head on the shoulders, and for fixing the larynx. The sternomastoid, the prominent muscle at the side of the neck, connects the sternum and clavicle below to the mastoid process of the temporal bone above.

THE BLOOD-VESSELS OF THE HEAD AND NECK

The Common Carotid Arteries

The common carotid arteries are the main arteries of the head and neck. They run upwards on either side of the neck, the right one coming from the brachiocephalic artery (a branch of the aorta), the left directly from the aorta. Each divides into: (a) an *internal carotid artery*, which passes upwards and through the base of the skull to supply the brain, and (b) an *external carotid artery*, which divides into branches which supply structures in the neck, face and scalp.

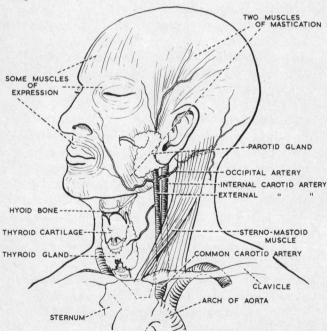

Fig. 5/3 Structures in the face and neck – oblique view

The Internal Jugular Veins

The internal jugular veins are the main veins in the neck. Beginning below the base of the skull as a continuation of a large venous sinus within the skull, each runs down the side of the neck with the internal and common carotid artery of the same side to join the subclavian vein to form a brachiocephalic vein.

The superficial vein visible in the neck of some people is the much smaller external jugular vein.

THE VERTEBRAL COLUMN

The vertebral column supports the head, neck and trunk and protects the spinal cord (see Fig. 5/4). It is composed of

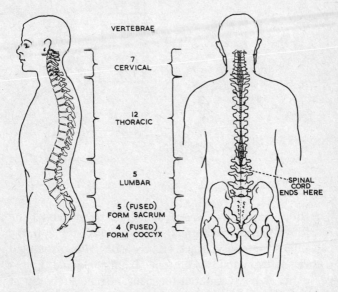

Fig. 5/4 The vertebral column – lateral view and posterior view

thirty-three bones and the ligaments that bind them together.

The bones are:

cervical vertebrae	7	in the neck
thoracic vertebrae	12	at the back of the chest
lumbar vertebrae	5	at the back of the abdomen
sacrum		at the back of the pelvis
coccyx		at the back of the pelvis

The vertebrae become progressively larger and stronger from above downwards as far as the upper end of the sacrum, where the weight of the body is transmitted through the sacro-iliac joints to the legs.

All the vertebrae, except the atlas, have a common shape: a body in front and an arch behind (see Fig. 5/5).

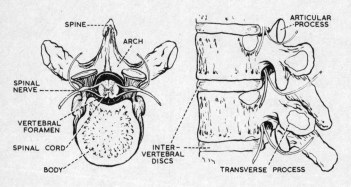

Fig. 5/5 Thoracic vertebrae, showing the spinal cord and nerves

The body is a thick mass, rounded in front; red bone marrow is present, in both childhood and adult life, in the cancellous bone in it. Each arch consists of two rounded pedicles in front and two flattened laminae behind. Transverse processes stick out at the sides, a spine sticks out at the back, and muscles acting on the vertebral column are attached to the processes and spine. Each vertebra articulates with the one above and the one below by means of articular processes. The body and arch enclose the vertebral foramen, a large hole, through which runs the spinal cord. The intervertebral foramina are small holes, through which pass the spinal nerves, at the sides of the vertebral column, between adjacent pedicles.

The Cervical Vertebrae

The seven cervical vertebrae are small and delicate. They differ from the others in having in each lateral process a hole through which passes the vertebral artery on its way to the brain. The *atlas*, the first cervical vertebra (named after the Greek giant who was supposed to bear the heavens on his head and hands), differs from the others in that it articulates with the occipital bone above and has no body, having in front only a ring of bone. The *axis*, the second cervical vertebra, has in the course of evolutionary development, taken over the body of the atlas, which sticks up as a peg-shaped process continuous with the body of the axis. This arrangement of the first two cervical vertebrae enables the head to be rocked on the neck at the joint between the occipital bone and the atlas and turned from side to side at the joint between the atlas and the axis.

The Thoracic Vertebrae

The twelve thoracic vertebrae are bigger and stronger than the cervical vertebrae. They become larger from above downwards, and the twelfth is a massive vertebra resembling a lumbar vertebra. The ribs articulate with their bodies and transverse processes, and the joint surfaces for these attachments distinguish the thoracic vertebrae from the others.

The Lumbar Vertebrae
The five lumbar vertebrae are massive bones with thick, strong spines and lateral processes.

The Sacrum and Coccyx
The *sacrum* is formed of five vertebrae fused together. It is wedge-shaped, being wide at the top and narrow at the bottom. It is concave in front and convex behind, and has holes for the spinal nerves to pass through. On either side is a large articular surface for the sacro-iliac joint, through which the weight of the body is transmitted to the legs. Below there is a small joint for the coccyx.

The *coccyx* is a little triangular bone, in which four degenerated vertebrae are fused together.

The Intervertebral Discs
The intervertebral discs are discs fixed to the bodies of adjacent vertebrae (see Fig. 5/5). Each disc is composed of an annulus fibrosus, a ring of fibrous tissue on the outside, and a nucleus pulposus, a jelly-like substance, on the inside. They act as shock-absorbers to the vertebral column and by allowing small movements to take place between adjacent vertebrae take part in the flexion and extension of the spine. In old age they can shrink and so reduce the height of a person.

The Vertebral Column
The vertebral (spinal) column is formed by the binding of the vertebrae into a functional whole by the intervertebral discs, by strong anterior and posterior ligaments running along the front and back of the bodies of the vertebrae, and by ligaments connecting adjacent spines and laminae. The vertebral column is not straight. As a baby lifts his head and later starts to walk, the column develops a bend forwards in

the neck, backwards in the upper thoracic region, forwards again in the lower thoracic and upper lumbar region, and ends in the curve of the sacrum and coccyx. A slight deviation (called a scoliosis) to right or left is often present, and curves have to be made to compensate for it above and below where it occurs.

The vertebral column is acted upon by powerful muscles in front and behind. Although individual vertebrae can move only slightly on one another, extensive movements can be made by the action of the vertebral column moving as a whole.

THE THORAX OR CHEST

The thorax is shaped like a cone, open above into the neck and closed off below by the diaphragm. The thoracic wall is composed of the thoracic vertebrae, the ribs, the costal cartilages and the sternum (see Fig. 5/6).

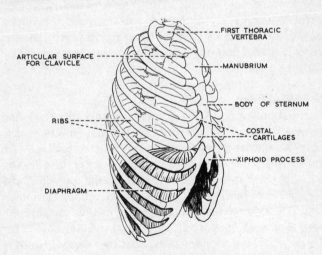

FIRST THORACIC VERTEBRA

ARTICULAR SURFACE FOR CLAVICLE

MANUBRIUM

BODY OF STERNUM

RIBS

COSTAL CARTILAGES

XIPHOID PROCESS

DIAPHRAGM

Fig. 5/6 The chest and the diaphragm

The Ribs

There are twelve ribs on each side. Articulating at the back with the bodies and transverse processes of the thoracic vertebrae, they sweep round the back and sides of the chest to end in the front of the thorax each in a piece of cartilage called a costal cartilage. The costal cartilages of the upper six ribs are attached directly to the sternum; the next four are attached each to the cartilage immediately above; and the bottom two, the costal cartilages of the short and imperfectly formed eleventh and twelfth ribs, are short, pointed and end in muscle. The intercostal spaces, the spaces between adjacent ribs, are filled by the external and internal intercostal muscles, which play a small part in breathing, pulling the ribs slightly outwards and upwards in inspiration and so slightly increasing the size of the thoracic cavity.

The Sternum (Breastbone)

The sternum is a flat bone in the middle of the front of the chest. It is composed of three parts: the upper wide manubrium, the long body, and the short xiphoid cartilage. The clavicle (collar bone) and upper costal cartilages articulate with it on each side.

The Diaphragm

The diaphragm is a dome-shaped structure of muscle and fibrous tissue between thorax and abdomen. Its muscle fibres arise from a circle of attachments – the bodies of the lumbar vertebrae at the back, the lower ribs at the side, and the xiphoid cartilage in front – and run upwards and inwards to be inserted into a flat central tendon of fibrous tissue. It is pierced by all the structures that run through both thorax and abdomen, the largest openings being for the aorta, the inferior vena cava and the oesophagus. It is the principal muscle of respiration. Its fibres are of striped voluntary muscle, and its movements are either automatic or voluntarily controlled.

THE PELVIS

The pelvis, a strong, massive ring of bone, at the lower end of the trunk, is composed of the two innominate bones at the front and the sides, and the sacrum and coccyx behind. It supports the abdominal contents and transmits the weight of the body to the legs (see Fig. 5/7).

The Innominate (Hip) Bone

The innominate bone is composed of three bones on each side: the pubic bone, the ischium and the ilium, which are separate bones up to puberty, when they start to fuse into one.

The *pubic bones* lie in front. They meet in the midline at the pubic symphysis, a slightly movable joint. Behind the pubic bone on each side is a large hole, the obturator foramen, which in life is filled in with fibrous tissue. The *ischium* is behind and below the pubic bone and part of it forms, deep in the gluteal region, the ischial tuberosity, a bony knob upon which we sit. The *ilium* forms a large flat wing of bone above the others, ending above in the long curve of the iliac crest, which extends from the anterior superior iliac spine in front to the posterior superior iliac spine behind. The *acetabulum* is the deep, rounded socket on the outer side of the innominate bone, where it articulates with the head of the femur to form the hip-joint. At the back the ilium articulates with the sacrum at the large, sacro-iliac joints.

The *true pelvis* is the lower part of the pelvis, bounded by the pubic, ischial, sacral and coccygeal bones. It contains the pelvic organs – the bladder, the rectum, and in women the internal genital organs. The female pelvis has a larger

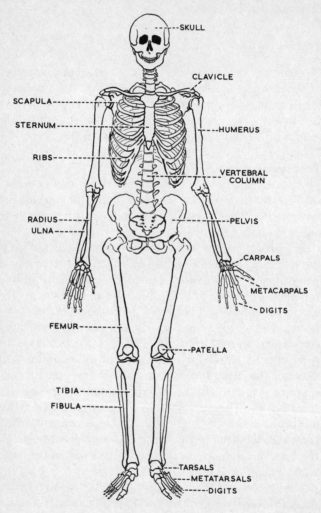

Fig. 5/7 Anterior view of the skeleton

cavity than the male pelvis, and is wider and shorter (see Fig. 5/8).

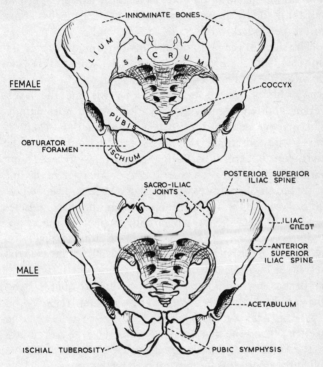

Fig. 5/8 The female and male pelvis

The Levator Ani

The levator ani is a cone-shaped muscle pointing downwards at the bottom of the pelvis. Its fibres are attached above to the inner surface of the true pelvis and running downwards and together support the bladder, the vagina in women and the rectum, and they close the opening of the pelvis below.

THE ABDOMINAL WALL

The abdominal wall extends from the costal margin (the border made by the lower ribs and costal cartilages) above to the pelvis below.

The sides of the abdominal wall are formed by three flat muscles – the external oblique muscle, whose fibres run downwards and forwards, the internal oblique muscle, whose fibres run upwards and forwards and the transversus abdominis muscle, whose fibres run horizontally. This arrangement of fibres gives both strength and suppleness to the abdominal wall; and a surgeon, operating on the abdomen, is careful as he makes his incision not to cut across the fibres of each layer but to divide each in the line of direction of its fibres and so damage it as little as possible. Towards the front of the abdomen the three muscles become fibrous sheets, which split to enclose the rectus abdominis muscle, a thicker muscle running vertically on either side of the middle line from the costal cartilages to the pubis (see Fig. 5/9). All these muscles maintain the abdominal organs in position and by contracting are able to bend the vertebral column forwards at the lumbar region.

At the back of the abdomen the space between the twelfth rib and the iliac crest is occupied by the beginnings of the three lateral abdominal muscles, by a flat muscle called the quadratus lumborum, and by the psoas muscle, a thicker muscle which arises from the bodies of the lumbar vertebrae and runs downwards into the thigh to be inserted into the upper end of the femur and is one of the flexors of the thigh on the abdomen.

THE ABDOMINAL CAVITY

The abdominal cavity is bounded above by the diaphragm, at the back by the lumbar vertebrae, whose bodies project

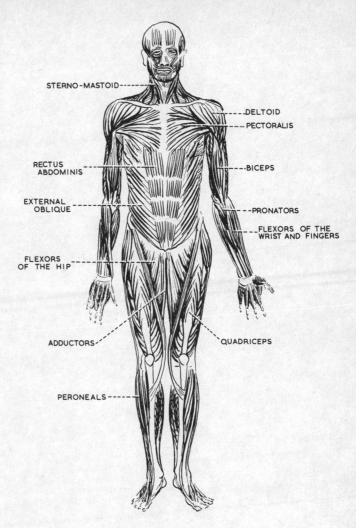

STERNO-MASTOID

DELTOID

PECTORALIS

RECTUS
ABDOMINIS

BICEPS

EXTERNAL
OBLIQUE

PRONATORS

FLEXORS OF THE
WRIST AND FINGERS

FLEXORS
OF THE HIP

ADDUCTORS

QUADRICEPS

PERONEALS

Fig. 5/9 The muscles of the body – anterior view

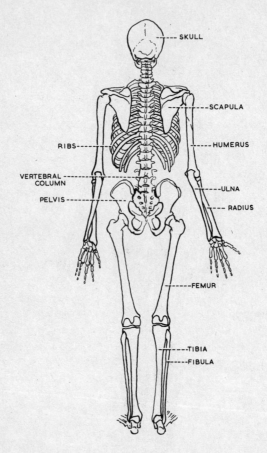

Fig. 5/10 The skeleton – posterior view

forwards into it, at the sides and front by the abdominal wall, and below by the pelvic floor.

It contains the stomach and the small and large intestines, the liver, the pancreas, the kidneys and the adrenal glands, the abdominal aorta and its branches, the inferior vena cava and its tributaries in the abdomen, the portal vein and its tributaries, and the autonomic nervous plexuses of the abdomen.

The pelvic part of the abdominal cavity contains
in both sexes: the bladder, the rectum and the anal canal;
in men: the seminal vesicles and the prostate gland.
and *in women*: the ovaries, the uterine tubes, the uterus and the vagina.

POSTURE

The upright posture of the body is maintained by the tone of many muscles. The *erector spinae* group of muscles is the most important of these. They are attached below to the sacrum and iliac crest, run up the back of the trunk, where they are attached to the vertebrae and the ribs, and at the top are inserted into the occipital bone of the skull. This group of muscles keeps the body and head vertical. The tone in its various parts is opposed by the tone of several muscles in front of the vertebral column in the neck, and by the tone of muscles in the front and sides of the abdominal wall (see Figs. 5/9 and 5/11).

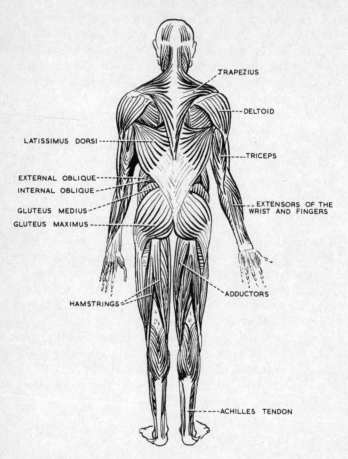

Fig. 5/11 The muscles of the body – posterior view

Chapter 6

ARMS AND LEGS

THE ARM

The Bones of the Arm

The shoulder-girdle is formed of two bones, the clavicle and the scapula (see Fig. 6/1).

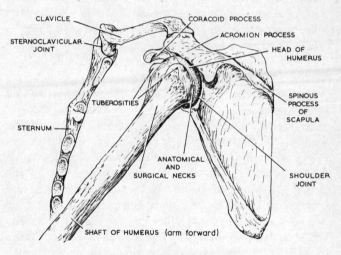

Fig. 6/1 The shoulder-girdle

The *clavicle* (collar-bone) is a thin bone which extends from the sternum (the sternoclavicular joint being the only articulation of the arm with the trunk) to the acromion process of the scapula, just above the shoulder-joint. Its function is to keep the shoulder-joint away from the chest wall so that the movements of the arm are not impeded. Of all the bones in the limbs it is the one most frequently broken, usually by a fall on to the shoulder.

The *scapula* (shoulder-blade) is a flat, thin bone, curved to fit the back of the chest wall, over which it moves with movements of the shoulder-joint. It has two processes sticking out of it: (a) a small coracoid process for muscular attachments, and (b) a larger spinous process, which arises from its posterior surface and ends in a thick acromion process above the shoulder-joint, articulating there with the outer end of the clavicle. The glenoid cavity of the scapula is a shallow depression which articulates with the head of the humerus to form the shoulder-joint. There is no joint between the scapula and the trunk, the scapula being attached only by muscles to the vertebral column, the chest wall and the humerus, an arrangement that allows it to have a wide range of movements over the chest wall.

The *humerus,* the bone of the upper arm, has a shaft and two ends, upper and lower. The upper end has a rounded head which articulates with the glenoid cavity of the scapula to form the shoulder-joint. The anatomical neck of the humerus is the constricted strip just below the head. The tuberosities are the two prominences (to which muscles are attached) just below the anatomical neck; and the surgical neck is the constricted part just below the tuberosities, being so-called because the bone is especially liable to break there (see Fig. 6/1). The shaft of the humerus is rounded in its upper part, but becomes flattened and wider towards its lower end. The lower end has articulatory surfaces for the radius and ulna, the two bones of the forearm. The medial and lateral epicondyles are the sideways projecting processes at the lower end of the bone.

The ulnar nerve runs in a groove in the back of the internal epicondyle, which has the name of 'funny bone' because of the peculiar sensation we get down the inner side of the forearm and the little finger side of the hand when we knock the ulnar nerve against the bone.

The *radius* and the *ulna*, the bones of the forearm, are long bones with shafts and upper and lower ends. The ulna, on the inner side when the palm is facing forwards, is the longer and articulates with the humerus by a notched process, of which the olecranon, the upper end, fits into a depression at the back of the lower end of the humerus, forming the prominent bone we can feel at the back of the elbow. The radius has at the upper end a rounded head, which articulates with the humerus above and the ulna on the inner side. Both bones are attached to one another for much of their length by an interosseous membrane of fibrous tissue, which helps bind them together, transmits forces from one to the other, and provides both in front and behind additional areas for the attachment of muscles of the forearm. The lower ends of both bones are thicker than the shafts above them, but each ends in a pointed styloid process. They articulate with each other and with the bones of the wrist. A Colles' fracture is a common fracture of the lower end of the shaft of the radius, the styloid process of the ulna often being broken off as well, as a result of a fall on to the outstretched hand.

The *wrist* or *carpus* comprises eight small bones of irregular shape arranged in two rows. In the palm of the hand are five *metacarpal bones*, one for each finger, the one for the thumb being the thickest and strongest. The *phalanges* (singular, *phalanx*) are the bones of the fingers; the fingers have each three, the thumb two.

The Shoulder-Joint
The shoulder-joint is a ball-and-socket joint between the head of the humerus and the glenoid cavity of the scapula. It has the characteristic features of a movable joint, with

cartilage covering the ends of the bones within the joint, a capsule of fibrous tissue, a synovial membrane, and ligaments strengthening the capsule on the outside. Special features of the joint are: (a) a fibrocartilaginous lip attached to the rim of the glenoid cavity deepens the joint; (b) the capsule is very loose, enabling the arm to be moved away from the side of the body, and (c) the acromioclavicular joint and the ligaments connecting the acromion with the coracoid process protect the joint above.

The shoulder-joint has a wider range of movement than any other joint. Its range is increased by simultaneous movements of the scapula on the chest wall. The arm can be moved forwards and backwards, away from or towards the body, can be rotated, or can be swung round the head in a circular sweep; and to enable all these movements to be made, the head of the humerus must be able to move freely in the shallow depression of the glenoid cavity. But, surrounded and strengthened by muscles in front and behind, the shoulder-joint is not easily dislocated by even a sudden jerk.

The Elbow-Joint

The elbow-joint is both a hinge-joint between the humerus above and the radius below, and a pivot-joint between the radius and the ulna (see Fig. 6/2). Both are enclosed in a common capsule, which is thickened and strengthened at the sides by ligaments running from the internal epicondyle to the ulna and from the external epicondyle to the radius. The head of the radius is held tightly against the ulna by the annular ligament, a ring of fibrous tissue.

At the elbow-joint the forearm can be extended or flexed on the upper arm, and the radius can be rotated upon the ulna so that it crosses over the ulna when the hand is turned palm backwards. Pronation is turning the palm backwards, supination is turning it forwards.

The elbow-joint can be dislocated backwards; and in a young child the head of the radius, before it is well developed, can be dislocated downwards through the annular ligament by a sudden jerk of the arm, as when a child is yanked out of the way of a car.

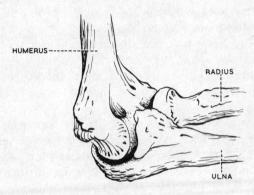

Fig. 6/2 The elbow-joint

The Wrist-Joint

The wrist-joint is formed above by the lower end of the radius and a fibrous disc over the lower end of the ulna and below by the first row of carpal bones. The articulation is a knuckle-like joint at which flexion (bending forwards), extension (bending backwards), adduction (bending to the little finger side) and abduction (bending to the thumb side) are possible. All these movements can be combined in a circular movement at the wrist.

The Muscles of the Arm

The main muscle-groups of the arm are:

(a) The pectoral muscles, which run from the front of the chest wall to the upper end of the humerus, lie deep to the breast, and form the front wall of the axilla (armpit) (see Fig. 5/9).

(b) The muscles at the back connecting the vertebral column and posterior chest wall to the shoulder-girdle and humerus (see Fig. 5/11).

(c) The deltoid muscle, which forms the shoulder.

(d) The muscles in the front of the upper arm, which flex the elbow and supinate the forearm. They include the powerful biceps muscle, whose strength has decided which way a corkscrew is made to turn.

(e) The muscles at the back of the upper arm, which extend the elbow-joint. The triceps muscle is the largest of them.

(f) The muscles in the front of the forearm, which are attached by tendons to the carpal bones and phalanges, flex the wrist and fingers, and pronate the hand.

(g) The muscles at the back of the forearm, which extend the wrist and fingers.

If we consider such a movement as grasping something, we can see that many muscles are called into play. The actual holding of an object in the hand is done by contraction of the flexor muscles of the forearm, aided by several small muscles in the hand. Simultaneously the extensor muscles at the back of the forearm have to relax, acting on the principle of reciprocal innervation. The wrist, the elbow and the shoulder have each to be held in an appropriate position by the action of other muscles.

Blood Vessels of the Arm

(a) *Arteries*

The *subclavian artery* supplies blood to the arm. The right subclavian artery arises directly from the brachiocephalic artery (a branch of the aorta) and the left directly from the aorta (see Fig. 6/3). It passes behind the clavicle and then into the axilla (armpit) where its name is changed to that of *axillary artery*. The axillary artery is continued down the arm as the *brachial artery*, which divides in front of the elbow into the *radial* and *ulnar arteries*, which supply the forearm and hand. The pulse is taken by feeling it in the radial artery at the wrist.

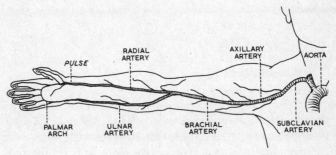

Fig. 6/3 The arteries of the arm

(b) *Veins*

The veins of the arm are superficial and deep. There are no superficial veins in the palms, for any veins there would be squashed out of existence by the grasping movements of the hand, and all the venous blood from the hand must pass into veins at the back of the hand, where they are not compressed. A large communicating vein is visible at the front of the elbow, and blood for examination is commonly obtained from it. The deep veins accompany the arteries. The veins eventually join to form the large subclavian vein on each side of the neck.

Lymph-Vessels of the Arm

The lymph-vessels of the arm pass superficially or with the arteries to the group of lymph-nodes in the axilla. The only other lymph-node in the arm is a small one in front of the internal epicondyle of the humerus.

Nerves of the Arm

The nerves to the skin and muscles of the arm come from the lower cervical and upper thoracic spinal nerves. They run behind the clavicle and in the axilla form the *brachial plexus*, a complicated criss-cross of nerves from which emerge the individual nerves to the arm.

THE LEG

The Bones of the Leg

The pelvic girdle is described on p. 49.

The *femur* (thigh-bone), the longest bone in the body, consists of a shaft and two ends (see Fig. 5/7 and 5/10). At its upper end it has a rounded head, which articulates with the acetabulum of the innominate bone to form the hip-joint. Immediately below the head is the neck of the femur, which in old age is likely to become devitalised, weak and liable to fracture. Just below the neck are two prominent trochanters to which muscles are attached. The shaft joins the neck at an angle of about 120° and runs downwards and inwards, surrounded by muscles and its back ridged by musculo-tendinous insertions. At the lower end the bone widens out into two large joint areas, the inner and outer condyles.

The *patella* (knee-cap) is a small detached bone, embedded in the tendon of the quadriceps muscle in front of the knee-joint. It is the largest example in the body of a *sesamoid bone*, which is a bone formed in a tendon where it passes over another bone.

The tibia and the fibula are the two bones in the leg between the knee and the ankle.

The *tibia*, the inner bone of the leg below the knee, has a broad, expanded upper end, which articulates with the condyles of the femur; a gradually tapering shaft, one surface of which can be felt under the skin of the shin; and a lower end which articulates with the talus at the ankle. The medial malleolus is its prominent projection at the inner side of the ankle.

The *fibula* is a long bone, attached at top and bottom to the tibia and running down the outer side of the leg. It is thin because it does not have to bear any weight, its functions being to act as a brace to the tibia and to provide surfaces for the attachments of muscles. The lower end articulates with the talus and forms the external malleolus. Like the radius and ulna, the tibia and fibula are attached to one another for most of their length by a fibrous interosseous membrane for the attachment of muscles. A Pott's fracture is a fracture at the lower ends of the tibia and fibula, and is partly a fracture, partly a dislocation, the fibula being broken just above the ankle and the medial malleolus of the tibia being broken or the ligament attached to it torn through and the foot pushed backwards.

The *tarsus* (the bones of the foot) is formed of seven bones: the *talus* helps to form the ankle-joint, the *calcaneum* is a strong thick bone which forms the heel and bears much of the weight of the body, and in front of these two are five smaller bones. In front of these bones are five *metatarsal bones*, one for each toe. The bones in the toes are called *phalanges*, like their counterparts in the fingers; all the toes have three, except the big toe which has two.

The Hip-Joint

The hip-joint is a ball and socket joint of great stability. It is formed by the articulation of the acetabulum of the innominate bone with the head of the femur (see Fig. 6/4). The head of the femur fits tightly into the acetabulum, whose depth is increased by a fibro-cartilaginous ring attached to its edge. A tight capsule is attached to this edge and to the neck of the femur. The capsule is strengthened by several ligaments, especially in front where a strong Y-shaped ligament attaches the ilium to the femur. A short ligament, called the ligamentum teres, runs from the acetabulum to a depression in the centre of the head of the femur and provides the means by which blood-vessels can get to the head.

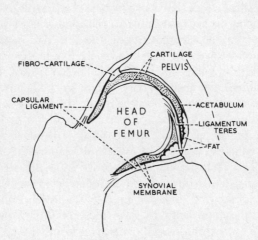

Fig. 6/4 The hip-joint

The movements possible at the hip-joint are flexion of the thigh on the trunk, where it is stopped by the thigh coming into contact with the abdominal wall; extension, any hyperextension being prevented by the Y-shaped ligament; abduction (movement away from the middle line of the body); adduction (movement towards it); and rotatory movements. The hip-joint is so strong that it is rarely dislocated, but a baby can be born with a congenital dislocation of the hip, apparently due to a congenital abnormality of the joint.

The Knee-Joint

The knee-joint is a hinge-joint, formed by the condyles of the femur, the upper end of the tibia and the patella (see Fig. 6/5). Except for the attachment of a ligament to its upper end, the fibula takes no part in the joint for it articulates with the tibia below the joint. The capsule of the joint is attached around the joint surfaces on the femur, tibia and patella. It is strengthened at the sides by strong ligaments, that on the outer side running from the external condyle of the femur to the head of the fibula and that on the inner side running from the medial condyle to the upper end of the tibia. It is strengthened in front by the broad tendon of the powerful quadriceps muscle, in whose tendon is embedded

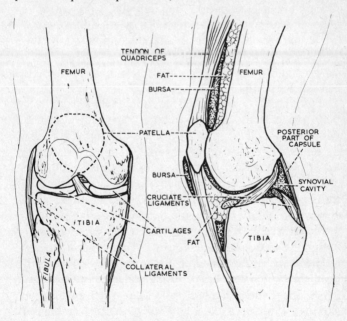

Fig. 6/5 The knee-joint

the patella. The cartilages of the joint (often torn by foot-ballers) are two half-moon shaped pieces of fibro-cartilage, resting on and attached to the top of the tibia; being thicker on the outer edge than the inner, they deepen the lower joint-surface. The cruciate ligaments are two strong fibrous cords which run across the middle of the joint, crossing as they go and helping to make the joint stable.

The movements at the knee are flexion, extension and a little rotation.

The Ankle-Joint

The ankle-joint is formed above and at the sides by the lower ends of the tibia and fibula and below by the talus. The talus is gripped between the two malleoli. The capsule of the joint is loose in front and behind to allow of movement and strengthened at the sides by strong ligaments.

The movement possible at the ankle-joint is the hinge movement used in walking.

The Arches of the Foot

The foot has to do two things: to support the weight of the body and to act as a lever in walking. A child is born with a flat foot, but arches are necessary for walking and the child develops them as he starts to walk. The arches of the foot are: (a) longitudinal, from front to back, the arch being much higher on the inner side than the outer, and (b) transverse, from side to side. The height of the arches varies very much in different people.

To withstand the weight of the body, which tends to flatten the arches, the foot is designed to maintain them (a) by the shape of the bones of the foot; (b) by short strong ligaments which connect the under surfaces of the bones; (c) by long tendons of muscles in the leg which are attached to bones in the feet and hold the whole foot as in a sling; (d) by short muscles in the sole of the foot, and (e) by the plantar fascia, a strong sheet of fibrous tissue, lying deep to the skin and connecting the front and back of the foot (see Fig. 6/6).

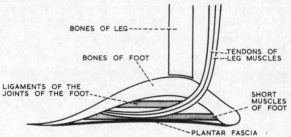

Fig. 6/6 Diagram of the structures forming the arch of the foot

The Muscles of the Leg

The main muscle-groups of the leg are:

(a) The flexors of the hip, which run from attachments to the front of the lumbar vertebrae and from the pelvis to the top of the femur.

(b) The gluteal muscles, which run from the back of the ilium and sacrum to the femur, form the mass of the buttock, and act on the hip-joint in various ways.

(c) The large quadriceps muscle in the front of the thigh, with its tendon inserted into the tibia and its action that of extending the knee.

(d) The adductor group at the inner side of the thigh, pulling the thigh inwards.

(e) The hamstring group at the back of the thigh, their tendons being inserted into the upper ends of the tibia and fibula; they flex the knee.

(f) The muscles in the front of the leg below the knee; they bend the foot and toes upwards.

(g) The peroneal group on the outer side of the leg; they steady the ankle and act on the foot.

(h) The calf muscles, ending in the strong Achilles tendon,

which is inserted into the back of the calcaneum; they pull the heel upwards in the action of walking.

(i) The deeper muscles in the calf, whose tendons run into the foot and bend the toes downwards.

(j) Small muscles in the foot which help to maintain the arches of the foot and act on the toes.

Walking

Walking is a complicated muscular action involving legs, arms and body.

If we analyse the movements made in taking a step forward, we find many muscle groups involved. Imagine that your right leg is about to move forward. Before you can begin to move it you must tilt the pelvis so that the weight of the body is transferred to your left leg. The right leg is then free to move. The heel is lifted off the ground by the action of the calf muscles. Next the knee is slightly flexed. The toes are momentarily left in contact with the ground, being used for pushing off, and then the whole leg is moved forwards by the action of the muscles acting on the hip, especially the powerful gluteal muscles. The heel is placed on the ground first, then the outer border of the foot, and finally the balls of the toes.

All this time the muscles acting on the trunk contract or relax to maintain an upright posture of the body, and the arms swing in helping to keep the balance.

Blood-Vessels of the Leg

(a) Arteries

The *femoral artery*, the main artery of the leg, is a continuation of the external iliac artery and enters the thigh at the middle of the groin. Having run down the inner side of the thigh, it passes to the back of the knee, where it becomes the *popliteal artery* which divides into an *anterior tibial artery* and a *posterior tibial artery*, which between them supply blood to the leg below the knee (see Fig. 6/7).

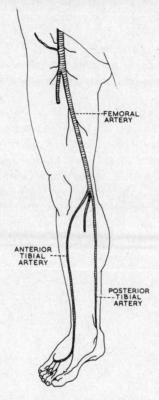

Fig. 6/7 The arteries of the leg

(b) *Veins*
The veins are both superficial and deep, the deep ones running with the arteries. They run into the large *femoral vein*, which lies on the inner side of the femoral artery and is continued in the abdomen as the external iliac vein. In

some people the veins in the legs become varicose, becoming larger than they should be and very tortuous, with valves so useless that the blood in the veins is almost stationary.

Lymph-Vessels of the Leg

Lymph-vessels pass up the leg to enter lymph-nodes at the back of the knee and in the groin.

Nerves of the Leg

The nerves to the leg come from the lumbar and sacral parts of the spinal cord. The *sciatic nerve*, the largest nerve in the body and about as broad as one's thumb, passes backwards out of the pelvis and then down the back of the thigh, giving off branches to the muscles and ending by dividing into two branches, which supply the leg and foot. The painful condition called sciatica is usually due to a prolapsed disc of the spinal column pressing on nerve-roots within the spinal canal.

Chapter 7

THE CIRCULATION OF THE BLOOD

The functions of the circulation are to carry to the tissues of the body the substances they require and to remove from them their secretions and the waste products they produce in their activities.

The cardio-vascular system, which performs these functions, consists of the heart, the arteries and arterioles, the thoroughfare vessels, the capillaries and sinusoids, and the venules and veins.

The basic actions are: the heart pumps the blood to the lungs, where it is oxygenated; from the lungs the blood returns to the heart; the heart then pumps it into the arteries and arterioles; from them it passes through either the thoroughfare vessels or the capillaries and sinusoids; thence it passes through the venules and is returned to the heart through the veins (see Fig. 7/1).

THE HEART

The heart lies inside the chest and between the lungs, and extends from the right side of the sternum to about 10cm from the middle line on the left side, where its apex can be seen and felt to beat in the space between the fifth and sixth ribs.

It is composed of three types of tissue:

1. The *endocardium* is the innermost layer and lines the outside of the valves and the smooth surface on the inner side of the chambers of the heart.

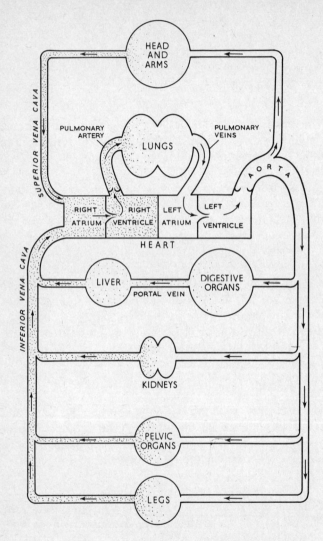

Fig. 7/1 A diagram of the circulation

2. The *myocardium* is the middle layer and is formed of the special cardiac muscle; it varies in thickness in different parts of the heart, being thickest in the left ventricle.

3. The *pericardium* is the outer layer and forms a double-layered fibrous bag, its two layers being continuous at the base (upper end) of the heart; adjacent surfaces of the bag are moistened with a little fluid so that one surface can glide smoothly over the other.

The heart is a pump and consists of two separate sides: a *right side*, whose function is to receive blood from the veins and pump it into the lungs, and a *left side*, whose function is to receive blood from the lungs and pump it round the rest of the body (see Fig. 7/2).

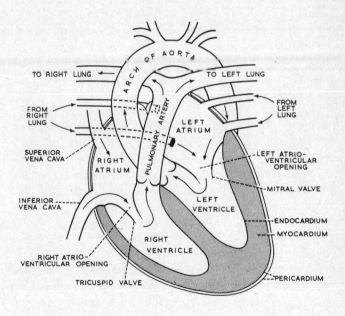

Fig. 7/2 A diagram showing the heart and main vessels

The Right Side of the Heart
The right side of the heart is composed of the right atrium and the right ventricle.

The *right atrium* is a thin-walled chamber into which open the superior and inferior venae cavae, the two large veins of the body. It communicates with the right ventricle through the right atrio-ventricular opening.

The *right ventricle* is situated next to the right atrium and has a moderately thick muscular wall because it has to pump the blood into the lungs. On its inner surface are numerous muscular bands and prominences, from which slim fibrous bands run to the edges of the tricuspid valve. Above it has an opening into the pulmonary artery.

The *tricuspid valve*, of three flaps of fibrous tissue covered by endocardium, surrounds the atrio-ventricular opening on the ventricular side and when it is closed prevents the blood in the ventricle from passing back into the atrium.

The *pulmonary valve* of three semilunar flaps, controls the opening of the ventricle into the pulmonary artery and prevents blood in the artery from falling back into the ventricle.

The *pulmonary artery* is a short wide artery leading out of the right ventricle and dividing into a right branch for the right lung and left branch for the left lung. The branches divide into smaller branches within the lungs.

The Left Side of the Heart
The left side of the heart is composed of the left atrium and the left ventricle.

The *left atrium* is a thin-walled chamber at the back of the heart. At the sides it receives the blood coming from the lungs through the pulmonary veins, two large veins on each side. Below it opens into the left ventricle through the left atrio-ventricular opening.

The *left ventricle* is situated below and in front of the left atrium and immediately to the left of the right ventricle, the two having a common wall. Its wall is about three times as thick as the wall of the right ventricle because it has the much harder work of pumping the blood to the most distant regions of the body. It has an opening above into the aorta.

The *mitral valve* (of two flaps like a bishop's mitre) guards the left atrio-ventricular opening, is fastened by several thin fibrous bands to muscular projections in the ventricular wall, and when closed prevents blood from passing back from the ventricle into the atrium.

The *aortic valve*, of three semilunar cusps, exactly like those of the pulmonary valve, surrounds the aortic opening and when closed prevents blood from passing back from the aorta into the ventricle.

The Cardiac Cycle

Cardiac muscle contracts and relaxes throughout life. The contraction of the heart is called its beat; and the cardiac cycle is the sequence of events that occur from the beginning of one beat to the beginning of the next. The heart beats at the rate of about seventy times a minute when a person is resting. Its rate is higher in infancy and when a person is excited, is taking exercise or has a raised temperature.

The *sinu-auricular node* is a small area of specialised cardiac muscle in the wall of the heart close to the entry of the superior vena cava. It is called the pacemaker of the heart because the heart-beat begins in it and it sets the rate of the heart. Having begun there each beat spreads through the two atria and then down the *atrio-ventricular bundle* or bundle of His, a narrow tract of special fibres, to get to the ventricles.

The two atria contract simultaneously and the two ventricles simultaneously and immediately after the atria, as soon as the impulse reaches them down the atrioventricular bundle.

Blood enters the right atrium through the superior and inferior venae cavae and is driven by atrial contraction into the right ventricle; and at the same time blood from the lungs is flowing into the left atrium through the pulmonary veins and is driven by atrial contraction into the left ventricle. Both ventricles then start to contract; the tricuspid and mitral valves close; the pulmonary and aortic valves open; the blood in the right ventricle is driven into the pulmonary artery and so into the lungs; the blood in the left ventricle is driven into the aorta and so into the rest of the body. Both ventricles then stop contracting and the pulmonary and aortic valves close. The atria fill with blood and the next beat begins.

Systole is the period during which the heart muscle is contracting. *Diastole* is the period immediately after systole during which the heart muscle is relaxed.

The heart produces several sounds during its beating, sounds which can be heard through a stethoscope or with the ear applied to the chest wall. The first heart sound sounds like 'lub' and is caused by the sudden tensing of the mitral and tricuspid valves at the beginning of ventricular systole. The second heart sound sounds like 'dub' and is produced by the closing of the aortic and pulmonary valves. Other sounds can sometimes be heard. In some diseases of the heart the sounds may be altered or other sounds called murmurs added to them.

Electrical changes taking place in the heart as a result of its muscular contraction can be magnified and recorded by an instrument called an electrocardiograph, the recording being called an *electrocardiogram* (ECG) (see Fig. 7/3). The electrical changes are recorded as waves labelled P,Q,R,S and T. The P wave is produced by contraction of the atria. Q,R and S waves are a complex produced by the contraction of the ventricles. T wave is produced by relaxation of the ventricles. Some diseases of the heart produce abnormalities of the ECG.

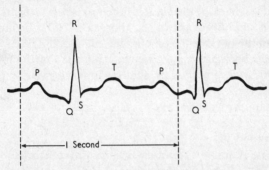

Fig. 7/3 A normal electrocardiogram, showing two beats of a heart beating about seventy times a minute.

The *right* and *left coronary arteries* supply the heart with blood. The blood in the chambers of the heart does not nourish the heart in any way. The two arteries arise from the aorta immediately above the aortic valve and run downwards over the heart, dividing into branches on its surface (see Fig. 8/2, p. 88). Their branches to the myocardium are particularly important. Disease of the coronary arteries causes angina pectoris, a severe pain in the chest on exertion, and a clot of blood in a diseased coronary artery can completely block it, causing a severe and often fatal coronary thrombosis.

THE ARTERIES AND ARTERIOLES

The arteries are tubes composed of a smooth inner layer called the intima, a middle layer of involuntary muscle fibres, and an outer coat of fibrous tissue.

The *aorta* is the largest artery of the body and from it are derived, directly or indirectly, all the other arteries of the body except the pulmonary artery, which passes from the right ventricle to the lungs. It begins at the aortic valve, passes at first upwards in the chest and then sweeps in an arch to the back of the chest. It then passes downwards in front of the lower part of the thoracic vertebral column, through the back of the diaphragm into the abdomen, and down the back of the abdomen as far as the fourth lumbar vertebra, where it ends by dividing into the two common iliac arteries.

Its main branches are:

Within the Chest:

The *right* and *left coronary arteries* to the heart.

The *brachiocephalic artery*, which divides into the *right common carotid artery* and the *right subclavian artery*. The common carotid artery supplies the neck, head and brain. The subclavian artery supplies the right arm; it gives off the *right vertebral artery*, which is one of the arteries to the brain.

The *left common carotid artery* to the neck, head and brain.

The *left subclavian artery* to the left arm and, through the *left vertebral artery*, the brain.

Within the Abdomen

The *coeliac artery* which supplies the stomach, the liver and the spleen.

The *superior mesenteric artery* which supplies the small intestine and part of the large intestine.

The *inferior mesenteric artery* which supplies the rest of the large intestine.

The *right* and *left renal arteries* to the kidneys.

The *right* and *left common iliac arteries*, each of which

divides into an *internal iliac artery*, which supplies the organs in the pelvis, and an *external iliac artery*, which passes into the thigh to become the *femoral artery*.

The Arterioles

The arterioles are the smaller branches of the arteries. They differ from the arteries in having a relatively thicker muscular coat. This muscular coat is normally in a state of slight contraction. The degree of contraction controls the amount of blood that enters an organ and is one of the factors in the production of the blood-pressure.

The arterioles divide into capillaries, sinusoids and thoroughfare vessels.

The Pulse and Blood-Pressure

The *pulse* is a wave transmitted along the arteries with each beat of the heart. It is not due to the passage of the blood, which travels much more slowly. It is most easily felt when an artery is pressed lightly against a bone, as at the wrist or in front of the ear.

The *blood-pressure* is the pressure exerted by the blood on the walls of the blood-vessels. It is the result of two forces: (a) the force of the heart's beat; and (b) the degree of contraction of the muscle of the arterioles. The pressure in the brachial artery is the one usually taken. It is taken with an instrument called a sphygmomanometer. Two measurements are taken: the *systolic pressure*, the pressure during systole, and the *diastolic pressure*, the pressure during diastole. Normal pressures in a young adult are: systolic 110–130mm of mercury, and diastolic 70–80 mm of mercury. The pressure usually rises a little with increasing years.

THOROUGHFARE VESSELS, CAPILLARIES AND SINUSOIDS

Thoroughfare vessels are thin-walled vessels which run directly from arteriole to venule. They are used for the rapid transmission of blood from the arteriole side to the venous side of the circulation.

Capillaries are thin-walled tubes formed of a continuous layer of cells. They run in the tissue-spaces between cells. Entrance to them is controlled by sphincters, which can open or shut according to the amount of blood required by a tissue at any one time; and when the sphincters are closed the blood is diverted through the thoroughfare vessels.

Gases and other chemical substances can pass through the capillary wall into the tissue-fluid in the tissue-spaces and so into the cells, and in the opposite direction from the tissue-fluid into the blood.

Sinusoids occur in the bone marrow, liver and endocrine glands. They are blood-channels without walls, where the blood is in direct contact with the cells of the organ and not separated from them by a capillary wall and tissue-fluid.

THE VENULES AND VEINS

The *venules* are small veins which collect the blood from the capillaries and conveys it to the veins.

The *veins* convey the blood from the venules to the heart. They are tubes composed, like the arteries, of an inner smooth coat, a muscular coat and an outer fibrous coat; but in them the muscular coat is much thinner than the muscular coat in the arteries, and the wall of a vein is therefore much thinner than that of an artery.

Many veins are provided with valves in their interior to prevent the blood from flowing in the wrong direction (see

Fig. 7/4). Each valve is a pocket-like flap of fibrous tissue covered with the inner coat and may occur singly or doubly. They produce a visible bulge in the vein. Some of these bulges can be seen in the veins of the forearm, especially if the onward flow of blood is prevented by compressing the arm, and if the blood is stroked through one of these valves and the finger then removed the blood will be seen to fall back as far as the valve and no further.

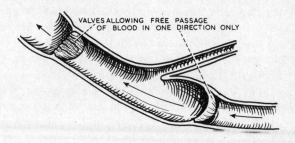

VALVES ALLOWING FREE PASSAGE OF BLOOD IN ONE DIRECTION ONLY

Fig. 7/4 Valves in a vein

Veins are usually superficial (just below the skin) or deep with the arteries. All the blood in them eventually passes into the superior vena cava or the inferior vena cava.

The *superior vena cava* is a single short vein just above and to the right of the heart. It is formed by the union of the right and left brachiocephalic veins and terminates in the right atrium. It transmits to the heart blood from the head and arms. The *inferior vena cava* is a single vein formed in the abdomen by the union of the two common iliac veins; it runs up the abdomen on the right side of the aorta, passes behind the liver and pierces the diaphragm to terminate in the right atrium. It transmits blood from the rest of the body.

Blood is moved along the veins by: (a) gravity in veins above the heart; (b) a muscle-pump of the muscles of the leg and abdomen; (c) the suction of the negative pressure in the chest during inspiration.

THE PORTAL SYSTEM OF VEINS

The portal system of veins enables the food-laden venous blood from the digestive tract to be transmitted directly to the liver.

It consists of veins from the stomach, intestines, spleen and pancreas. The *portal vein* is a vein formed by the union of veins from these organs. It is a short vein which passes to the back of the liver and there divides into branches. These branches open into the sinusoids of the liver, the blood being brought into direct contact with the liver cells (see Fig. 12/1, p. 122). From these sinusoids the blood passes into two hepatic veins, which enter the inferior vena cava at the back of the liver, the blood being returned in this way into the general circulation.

Chapter 8

RESPIRATION

Respiration consists of the exchange of gases in the lungs, the transport of gases in the blood, and the exchange of gases between blood and tissues. The gases are oxygen and carbon dioxide. Oxygen, which is necessary for all vital processes, has to be taken out of the air into the body, and carbon dioxide, a waste-product of cellular activity, has to be removed from the body.

THE RESPIRATORY PASSAGES

The respiratory passages are the passages through which the air has to pass as it is breathed in and out of the lungs. They are: the nose, the pharynx, the larynx, the trachea, and the bronchi (see Fig. 8/1).

Except for part of the larynx they are lined throughout with ciliated cells whose function is to move any dust away from the lungs.

The Nose

The nose consists of the external organ and the nasal cavities behind it. The two nasal cavities are separated by a midline septum (often deviated to one side or the other) composed of bone and cartilage. Three small turbinate bones are attached to the outer wall of each cavity and project into it. The accessory air sinuses – cavities in the maxillary, frontal, ethmoid and sphenoid bones – communicate through holes with the nasal cavities. The whole of the interior of the cavities, the turbinate bones and the air sinuses are lined with a continuous vascular mucous membrane.

Hairs in the nostrils filter off gross particles of dust, and the mucous membrane warms and moistens the air before it enters the lungs.

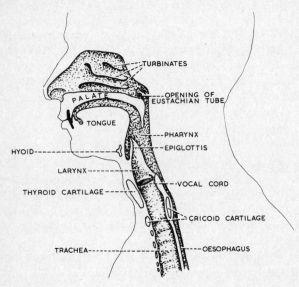

Fig. 8/1 A section through the respiratory passages

The Pharynx
The *pharynx* is the tube behind the nose and mouth and in front of the cervical spinal column. The air that enters through the nose and mouth passes through it on its way to the larynx.

The Larynx
The *larynx* (voice box, Adam's apple) is formed by several pieces of cartilage: (a) the thyroid cartilage, large and

V-shaped; (b) the epiglottis, projecting upwards under the tongue; (c) the cricoid cartilage, a ring of cartilage below the thyroid cartilage, and (d) the arytenoid cartilages, two tiny pyramids perched on the back of the cricoid cartilage and projecting upwards within the V of the thyroid cartilage.

The *vocal cords* are two slender bands of fibrous tissue stretched between the thyroid cartilage in front and the arytenoid cartilages behind. All the air passing through the larynx has to pass between the vocal cords, which can be moved apart or closer by the action of some small muscles attached to the arytenoid cartilages. The interior of the larynx is lined with mucous membrane, except for the vocal cords which are covered with a thinner membrane because in speaking they have to move at great speed.

The Trachea and Bronchi

The *trachea* (windpipe) is a tube about 10cm long, which extends from the bottom of the larynx into the upper part of the chest where it divides into a right and left bronchus (see Fig. 8/2). It is made of rings of cartilage, incomplete at the back, the space there and the spaces between rings being filled with fibrous and elastic tissue. This kind of construction is like corrugated piping and prevents the trachea from becoming kinked when the neck is bent or turned.

The right and left *bronchus* pass from the trachea to the root of the right and left lung respectively. They are made of the same material as the trachea, but they are narrower and their cartilaginous rings are thinner. At the root of the lung each bronchus divides into smaller branches.

THE LUNGS

The right lung is composed of three lobes – upper, middle and lower; the left is composed of two lobes – upper and

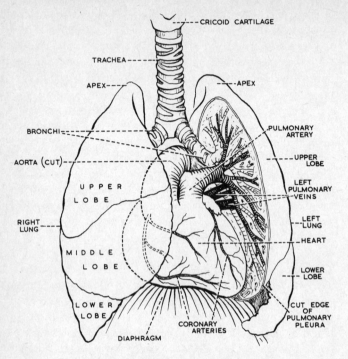

Fig. 8/2 The heart and lungs

lower. The left lung is thinner than the right because of the presence of the heart on the left side. The apex of each lung reaches into the neck just above the clavicle; its base rests on the diaphragm; its outer surface is pressed against the chest wall; its inner surface is moulded around the heart. The root of the lung is the place on its inner surface where structures enter or leave it. These structures are the bronchus, the pulmonary artery, the pulmonary veins, lymph-vessels and nerves.

The *pleura* is the membrane which encloses each lung. It is a double-layered bag. Its inner layer is firmly attached to the surface of the lung; its outer layer lines the chest wall; and the two layers are continuous around the root of the lung. The adjacent surfaces of the pleura are moistened with a little fluid, which enables them to move easily over each other in the movements of respiration.

Each lung is composed of a number of segments, each segment being a self-contained unit with its own blood supply and main branch of a bronchus. Within its segment this branch divides into smaller branches called bronchioles (see Fig. 8/3). A *bronchiole* has no rings of cartilage in it, but it has a ring of muscle which by contracting and relaxing can alter the size of the lumen of the tube. The smallest bronchi end by opening into alveoli. An *alveolus* is an air-containing sac with a wall only one cell thick and surrounded by a plexus of capillaries.

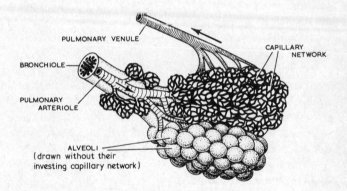

Fig. 8/3 The circulation of blood over the alveoli of the lungs

The tissues of the lungs are normally soft, spongy and filled with air. In new-born babies they are pink, but they become darker with age as carbon is deposited in them, especially in smoke-breathing town dwellers.

RESPIRATION

Respiratory Movements

The movements of respiration are inspiration and expiration (see Fig. 8/4). Normally at rest these movements occur about sixteen times a minute. There is a slight pause after expiration. The movements are controlled through small nerve centres: the *respiratory centre* in the medulla oblongata of the brain, and the two *carotid bodies*, small structures on each side of the neck, situated just where the common carotid artery divides into its two branches.

The cells in the centres are sensitive to the amount of carbon dioxide in the blood passing through them. If the amount of carbon dioxide is increased, nervous impulses from them produce deeper and faster breathing. If the amount of carbon dioxide is decreased, a reduction of nervous stimulation produces slower and shallower breathing. Lack of oxygen will stimulate the centres, but to a much smaller degree.

In *inspiration* contraction of the muscle of the diaphragm pulls down its central tendon, contraction of the intercostal muscles turn the ribs slightly outwards and lifts them up a little, and the sternum is moved slightly forwards. All these movements increase the volume of the thorax, which the lungs, under the pressure of the atmosphere, expand to fill. In *expiration* the muscle fibres of the diaphragm relax, allowing the central tendon to rise to its resting position, the ribs and sternum fall inwards, and air is driven out of the lungs. In inspiration and expiration there is much more movement at the bases than at the apices of the lungs.

Oxygen and Carbon Dioxide

Air is composed of oxygen, carbon dioxide, nitrogen, some of the inert gases (such as argon) and a variable amount of

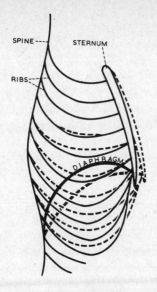

Fig. 8/4 The chest wall in inspiration and expiration
---- = inspiration
—— = expiration

water vapour. As oxygen is absorbed into the blood from the inspired air and carbon dioxide is excreted in the expired air, inspired air and expired air contain different proportions of them.

	Inspired Air	Expired Air
Oxygen	21.00 per cent	16.00 per cent
Carbon dioxide	0.04 per cent	4.00 per cent

Nitrogen, the largest constituent of air, is not affected by respiration, being breathed in and out of the lungs without being absorbed or altered in any way. About 400ml of air are breathed in and out in a single quiet respiration, but some of this does not reach the alveoli but remains in the 'dead space' of the respiratory passages where no gaseous exchanges can take place.

The dark venous blood enters the lungs through the pulmonary artery and its branches. As the blood passes through the network of pulmonary capillaries over the alveoli, oxygen passes from the air into the blood and carbon dioxide from the blood into the air. The oxygenated and now bright red blood returns to the heart through the pulmonary veins.

Oxygen is transported in the blood partly in the red cells as *oxyhaemoglobin*, the oxygen being combined with haemoglobin in the cells, partly dissolved in the plasma of the blood. It is the red colour of oxyhaemoglobin which gives to arterial blood its red colour. As blood circulates round the body, the process of internal or tissue respiration takes place. The combination of oxygen with haemoglobin being a weak one, oxygen is freely given up to the tissues that require it, the oxygen passing out of the blood into the tissue-spaces and then into the cells themselves. Simultaneously the carbon dioxide produced by cellular activity passes from the cells into the blood, which conveys it, partly in the plasma, partly in red cells, to the lungs for excretion. An increase in the amount of carbon dioxide in the blood stimulates the respiratory centres to increase the depth and rate of respiration in order to reduce it.

Chapter 9

THE BLOOD

Blood is a thick and opaque fluid, kept in constant circulation through the heart and blood vessels. The amount in a man is about 5 litres, in a woman about 4 litres. It is the transport medium of the body. It carries food and oxygen to the tissues of the body and removes carbon dioxide and other waste products of cellular activity. In addition, it helps to maintain constant the temperature of the body and plays an important part in waging war on harmful microorganisms and in the establishment of immunity to infection.

It is composed of plasma, red cells, white cells and platelets.

THE PLASMA

The plasma is a clear yellow fluid which forms just over half the total amount of the blood. It is slightly alkaline and its chemical composition is kept as constant as possible. It is composed of:

1. *Water* which forms about 92 per cent of plasma.
2. *Proteins* which include globulin, fibrinogen, prothrombin and antigens. They are involved in transporting salts, vitamins, etc. around the body, in maintaining the reaction of the plasma, in the clotting of the blood, and in the development of immunity.
3. *Amino-acids* which are absorbed into the blood as the end-products of protein digestion. They pass in the

blood of the portal veins to the liver and thence to the rest of the body. They are ultimately broken down in the liver.

4. *Urea* which is an end-product of amino-acid metabolism. It is formed in the liver, transported in the blood to the kidneys and excreted in the urine.

5. *Fats* which are transported in the blood after absorption from the small intestine and from the fat-depots of the body, and are broken down into simpler substances in the liver.

6. *Glucose* which is present in amounts which vary with the absorption of sugar from the small intestine and its utilisation by the cells of the body.

7. *Salts* which in the plasma include chlorides, phosphates, sodium, calcium and potassium salts.

8. *Hormones* which are the secretions of the endocrine glands and are present in the plasma in minute amounts.

THE CELLS OF THE BLOOD

Red Blood Cells

A red blood cell (erythrocyte) is strictly speaking not a cell because it does not have a nucleus, and it is therefore sometimes called a corpuscle and not a cell.

The numbers normally present are: (a) in a man, 5 million per mm^3 of blood, and (b) in a woman, 4.5 million per mm^3 of blood.

A red cell is a circular, biconcave disc (see Fig. 9/1). It contains haemoglobin within a thin membrane. Haemoglobin is an iron-containing pigment, which combines with oxygen to form oxyhaemoglobin. Its principal function is the transport of oxygen to the tissues from the lungs.

The substances necessary for the formation of red cells

are iron, vitamin B_{12}, folic acid, (a member of the vitamin B group) and, traces of cobalt and copper.

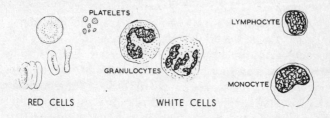

Fig. 9/1   Red and white blood cells and platelets

Before birth, red cells are formed from nucleated cells in the spleen, liver and bone marrow, but after birth they are formed in bone marrow only. They lose their nuclei just before they pass into the bloodstream, and the appearance of nucleated cells in the blood is an indication of hasty blood formation, as happens after a severe haemorrhage. Red cells live for about 120 days and are then broken down in reticulo-endothelial cells in the liver, spleen and bone marrow. The iron from them is retained in the body and used to make new haemoglobin. The rest of it is converted into other substances and excreted in the bile.

White Blood Cells

There are three kinds of white cells, granulocytes, monocytes and, lymphocytes (see Fig. 9/1). The total number of white cells is 4 000–11 000 per mm³ of blood.

Granulocytes are often called polymorphs (shortened from polymorphonuclear cells) because of the different shapes the nucleus can assume, being in some horseshoe-shaped and in others partly divided into two, three or more lobes – the older the cell the more lobes it has. Each granulocyte is about half as big again as a red cell. Their cytoplasm contains many tiny granules. By their staining reactions granulocytes can be divided into:

neutrophils: 2 500–7 500 per mm³; granules do not stain

eosinophils: 150–400 per mm³; granules stain red

basophils: 0–100 per mm³; granules stain blue.

Granulocytes are formed in bone marrow, live for a few weeks, and are then destroyed by cells of the reticulo-endothelial system. They can move with amoeboid-like movements. The neutrophils attack micro-organisms that have invaded the body. Such an invasion stimulates the production of neutrophils in large numbers. They are attracted towards the micro-organisms, and to get at them they squeeze through the capillary walls, and by an action called phagocytosis surround them, engulf them and destroy them.

The function of the eosinophils is not definitely known. Basophils may be involved in the transport of histamine, a chemical substance necessary for certain reactions.

Monocytes are large cells, nearly twice as big as the granulocytes. Their function is a phagocytic one similar to that of the neutrophils.

Lymphocytes are involved in immunity reactions (see Chapter 20).

Platelets

Platelets are tiny, round or oval discs without nuclei. They are derived from large cells in the bone marrow. There are 250 000–500 000 per mm³ of blood.

In the bloodstream they flow along close to the walls of blood-vessels, where they can carry out their functions of maintaining intact the inner layer of the vessels and producing thromboplastin, which is necessary for the clotting of blood after a haemorrhage.

THE CLOTTING OF BLOOD

When blood is shed it clots in about three minutes into a jelly-like substance, the actual time taken being determined mainly by the temperature of the air. The chemical processes that cause blood to clot are complicated, involving at least twelve substances present in plasma. Among them must be: prothrombin, calcium, thromboplastin (derived from platelets and damaged tissues), fibrinogen, and vitamin K.

By the interactions of these twelve substances, fibrin is formed and makes a web in which the red cells become entangled and this becomes a clot. Fibrin contracts in time and squeezes *serum*, a pale yellow fluid, out of the clot.

THE BLOOD-GROUPS

Every person's blood belongs to one of four main blood groups – AB, A, B, and O. The blood of one person is said to be incompatible with the blood of another person when mixing the blood causes the red cells to clump together, a clumping that can be seen with the naked eye and looks like brick dust. The incompatibility is due to interaction between certain chemical substances present or absent on a person's red cells and in his plasma.

The proportion of people in each group is:

group AB	5 per cent
group A	40 per cent
group B	10 per cent
group O	45 per cent

When a person is given a blood transfusion, he must be given blood of the same group as himself.

Rhesus Factor

This factor gets its name because it was first found in the Rhesus monkey.

About 85 per cent of people have the Rhesus antigen and are called Rh-positive and the other 15 per cent do not have it and are called Rh-negative. It is important that a female who is Rh-negative is never given a transfusion of Rh-positive blood for any child born to her might be seriously affected by the changes it can produce. Serious effects can follow the conception of a Rh-positive child by a Rh-positive father in a Rh-negative woman, and if the woman is not given appropriate treatment just after the birth of the child any subsequent child might develop severe anaemia and jaundice owing to destruction of its red blood cells.

A Rh-negative male should not be given Rh-positive blood because he would develop antibodies which would be dangerous if he happened to be given a second transfusion of Rh-positive blood.

THE SPLEEN

The spleen is a soft, purplish-red organ about 13cm long, 5cm broad and 2.5cm thick. It lies in the left upper region of the abdomen, under cover of the ribs and against the under surface of the diaphragm (see Fig 10/4, p. 106). It is composed of a pulp of connective tissue, red and white blood cells, sinusoids and lymph-follicles (patches of lymph-tissue identical with that in lymph-nodes), all enclosed in a thick fibrous capsule. It is connected to the rest of the body by an artery and vein, and its size varies with the amount of blood in it.

Its functions are: (a) to form red blood cells (in fetal life only); (b) to destroy old red blood cells; (c) to be a store of iron, derived from old red cells; (d) to be a store of platelets; (e) to form lymphocytes in its lymph follicles (f) to form immuno-globulins (necessary for the development of immunity); and (g) to filter foreign particles out of the blood.

THE RETICULO-ENDOTHELIAL SYSTEM

The reticulo-endothelial system is composed of cells which do not form one organ or set of organs, but are present in organs and tissues in various parts of the body. They are found in spleen, lymph-nodes, liver, thymus, bone marrow and blood-vessels. The principal functions of the cells of this system are the removal of particles of foreign matter and dead cells and the destruction of old red cells.

Chapter 10

THE ALIMENTARY TRACT

The alimentary tract is in essence a long tube with the mouth at one end and the anus at the other, and between the two extremities mechanisms by which food is broken down into forms small enough to be absorbed through the wall of the intestine. This breaking down is done partly by chemical action, partly by physical movements. Much of it is done by enzymes – catalysts formed by the glands that open into the alimentary tract (a catalyst is a substance that produces chemical changes in another substance without itself being changed).

The alimentary tract consists of the mouth and pharynx, the oesophagus, the stomach, the small intestine and the large intestine.

THE MOUTH AND PHARYNX

The *mouth* is formed by the lips in front, the cheeks at the side, the palate above and the tongue below. Behind it opens into the oral part of the pharynx (see Fig. 8/1, p. 86). The *pharynx* is continuous with the nasal pharynx above and opens below into the larynx in front and the oesophagus behind.

The *tongue* is composed of muscle attached to the mandible and hyoid bone. It is both an organ of taste, with taste buds on its surface, and a muscular organ involved in mastication, swallowing and speech. Solid food is broken up in the mouth by chewing movements of the jaws assisted by movements of the tongue. The incisor teeth are shaped for biting food, the premolars and molars for crushing it. As it is chewed it is mixed with saliva so that it can be easily swallowed.

The *saliva* is the secretion of three pairs of glands – the parotid, submandibular and sublingual glands. The parotid glands are below and in front of the ears; the submandibular glands are between the mandible and the tongue; the sublingual glands are below the tongue. Saliva passes through ducts from the glands to the inside of the mouth. It is a thick, colourless and slightly alkaline fluid which contains ptyalin, an enzyme which acts on starch. Saliva is necessary for the chewing of food to prepare it into a soft lump that can be swallowed, to begin the digestion of starch, to keep the mouth, tongue and teeth clean, and for speech.

Swallowing
Swallowing is a rapid and complex action in which food is thrown from the back of the tongue, through the pharynx and into the top of the oesophagus (see Fig. 10/1). The

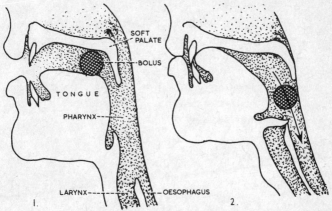

Fig. 10/1 The mechanism of swallowing

difficulty is in getting the food from the tongue to the oesophagus without letting it go down the larynx, which is just below the tongue. Food in the mouth is rolled into a soft ball called a bolus, and this bolus is pushed on to the back of the tongue. All the openings into the mouth are then closed except that into the pharynx. The lips and jaws are shut, the soft palate moves upwards to shut off the way into the nasopharynx, the tongue moves upwards and back-wards, the larynx is lifted under the tongue, which now projects over it, and the vocal cords are brought together. The way is now safe. With a contraction of the tongue the bolus is thrown into the pharynx and falls into the opening of the oesophagus. A little air is swallowed with each bolus.

THE OESOPHAGUS

The oesophagus is a tube about 25cm long. It begins at the lower end of the pharynx, runs down the neck behind the larynx and trachea and down the thoracic cavity behind the heart; it then pierces the diaphragm and enters the stomach (see Fig. 10/2). It has an inner coat of mucous membrane, a middle coat of muscle, and an outer coat of fibrous tissue.

When a bolus of food enters the oesophagus, peristaltic waves start to run down the oesophagus, and the bolus goes down the tube partly under the influence of gravity, partly squeezed down by the wave. A peristaltic wave is a wave of muscular relaxation followed by a wave of muscular con-traction. The lower end of the oesophagus relaxes to let the bolus into the stomach.

THE STOMACH

The stomach is the most dilated part of the alimentary tract. It lies in the upper half of the abdomen and usually is J-shaped, but when full it falls lower in the abdomen and its

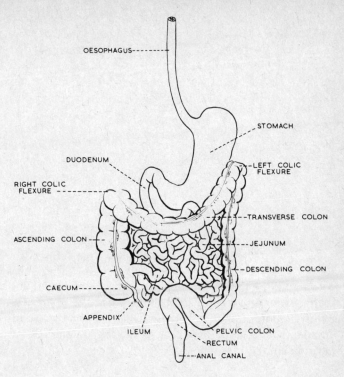

Fig. 10/2 The digestive tract

shape can vary with the amount of food in it and the position of the body (see Fig. 10/3).

It is like a bag with the left side, the greater curvature, much longer than the right side, the lesser curvature. The oesophagus enters it just below the upper end at an opening called the cardiac opening. The body of the stomach is its largest part. The pyloric canal is the narrow lower part of the stomach; it opens into the duodenum through the pyloric opening, which is surrounded by the pyloric sphincter, a thick ring of muscle. The wall of the stomach is

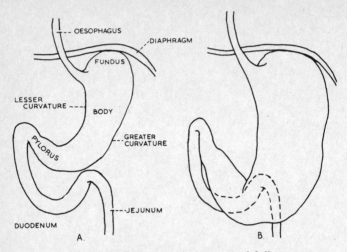

Fig. 10/3 The stomach, empty and full

composed of an inner mucous membrane, which contains many glands and is thrown into folds when the stomach is empty, a submucous coat of loose tissue, a muscular coat of circular and longitudinal fibres, and an outer coat of peritoneum.

Gastric Juice

Gastric juice is the fluid secreted by glands in the mucous membrane of the stomach. It is produced partly as a reflex reaction to the sight, smell and taste of food and partly as a reaction to chewing, swallowing and the arrival of food, especially meat, in the stomach. It is acid because it contains hydrochloric acid. It also contains pepsin, an enzyme, and mucinogen, which coats the surface of the mucous membrane with mucus to prevent the mucous membrane itself from being digested by gastric juice.

Gastric juice starts the breaking down of protein, halts the action of the alkaline ptyalin of the saliva, turns milk into a light clot, kills some micro-organisms or prevents them from multiplying.

Gastric Movements
The stomach does not move much when it is empty. While there is food in the stomach, it is kept moving by peristaltic waves. These begin at the upper end, increase in size as they pass down the body of the stomach, and fade away as they reach the pyloric canal. At first the pyloric sphincter remains closed, and the food is forced up and down in the stomach. After a time the pyloric sphincter relaxes and then the peristaltic waves start to drive the food into the duodenum. After a meal the stomach normally empties within five hours. Much fat in a meal delays emptying. Excitement will hurry a meal through, fear delay it.

THE SMALL INTESTINE

The small intestine extends from the stomach to the large intestine and is about 6.5 metres long, and to be contained within the abdomen it has to be coiled up in many coils. It consists of the duodenum and the jejunum and ileum (see Fig. 10/2).

The Duodenum
The duodenum is the first part of the small intestine and is about 25cm long (see Fig. 10/4). It begins at the pyloric opening of the stomach and is curled in a big C around the head of the pancreas. The first part of it can move a little with the stomach, but the rest is fixed to the back of the abdomen in front of the right kidney and the inferior vena cava. The bile duct and the pancreatic duct pass through the wall of the descending part of the curve and open, usually at an opening common to both of them, into the interior of the duodenum.

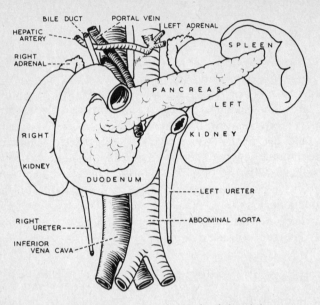

Fig. 10/4 The structures at the back of the abdomen

The Jejunum and Ileum

The jejunum and ileum are the second and third parts of the small intestine, the jejunum forming the first two-fifths and the ileum the next three-fifths, but there is essentially no difference between the two parts and the names are traditional. The jejunum begins at the end of the duodenum and lies mainly in the middle of the abdomen. It is continuous with the ileum, which lies mainly in the lower half of the abdomen and ends by joining the large intestine at the ileo-caecal valve.

The walls of the small intestine are composed of a mucous membrane, a submucous coat, a muscular coat of circular and longitudinal fibres, and a peritoneal coat. The mucous membrane has many circular or semicircular folds and its surface is marked by millions of villi, tiny moving projections which make it look like velvet. The surface of the mucous membrane shows Peyer's patches, which are patches of lymph tissue and are more common and prominent in the ileum than the jejunum.

Intestinal Movements

Intestinal movements begin with the entrance of partly digested food into the duodenum. They take the form of small, irregular contractions of a segment of intestine, which pass along it for a little way and then fade out. They are sufficiently powerful to keep the intestinal contents moving along.

Chemical Action in the Small Intestine

The intestinal contents are acted on by three fluids: (a) the *intestinal juice*, secreted from glands in the mucous membrane of the small intestine and containing enzymes which break down proteins, fats and carbohydrates; (b) the *bile*, whose salts make fat particles smaller; and (c) the *pancreatic juice*, which contains enzymes which break down proteins, fats and carbohydrates.

Digestion in the Small Intestine

Chyme is the semi-liquid substance produced by the action of gastric digestion on food. As it enters the duodenum it consists of water, of proteins, fats and carbohydrates in a partly digested state, vitamins, minerals and cellulose. In the small intestine enzymes break the proteins down to amino-acids, carbohydrates into single-sugars, and fats into fatty acids and glycerol. All these products of digestion are then in forms capable of passing through the mucous membrane of the small intestine and into the capillaries of the portal system of veins, through which they are transported to the liver. Some particles of fat are not broken down, and passing into the lymph vessels (lacteals) of the intestine they are conveyed along the thoracic duct and eventually into the blood and so into the liver. The vitamins and minerals are absorbed through the intestinal wall into the portal capillaries. Cellulose is unchanged.

By the time the intestinal contents reach the ileo-caecal valve all the products of the digestion of protein, fat and carbohydrate have been absorbed, and the intestinal contents have become a bile-stained fluid containing water, cellulose and cells cast off from the intestinal mucous membrane.

THE LARGE INTESTINE

The large intestine extends from the ileo-caecal valve to the anus and is about 1.5 metres long (see Fig. 10/2). Its walls are composed of a mucous coat, smooth and without villi, a submucous coat, a muscular coat and an outer coat of peritoneum. It differs from the small intestine in being larger and more fixed in position, in having little nodules of fat attached to the outside of much of it, and in having its longitudinal muscle fibres arranged in three visible bands called the taeniae coli, which being short give the large intestine a puckered appearance. It consists of the caecum, the appendix, the colon, rectum and the anal canal.

The Caecum

The caecum is a short wide sac and lies in the right iliac fossa. The ileum opens into its left side through an opening controlled by the ileo-caecal valve, which allows intestinal contents to pass from ileum into caecum but not in the opposite direction. The opening of the appendix is just below the valve. Above the caecum is continuous with the colon.

The Appendix

The appendix is a thin, worm-like tube up to 18cm long. It opens above into the caecum and below ends in a blind tip. Its position is variable: usually it lies behind the caecum, but it can point towards the pelvis or lie in front of or behind the end of the ileum. Its wall contains much lymph tissue.

The Colon

The colon is composed of four parts: ascending, transverse, descending and pelvic. The *ascending colon* runs upwards from the caecum to the under surface of the liver, where it turns to the left and forwards in a sharp bend, the right colic flexure. The *transverse colon* begins at the right colic flexure, loops across the abdomen from right to left, and climbs to below the spleen, where it bends abruptly at the left colic flexure. The *descending colon* begins at the left colic flexure, runs down the left side of the abdomen, curves inwards towards the midline and ends at the brim of the true pelvis. The *pelvic (sigmoid) colon* begins at the brim of the true pelvis, forms several loops within the pelvis, and ends in the rectum.

The Rectum and Anal Canal

The *rectum* is about 12cm long and ends at the anal canal. It lies in front of the sacrum and coccyx and behind the bladder and prostate gland in the male and behind the uterus and vagina in the female.

The *anal canal* is short and narrow and ends at the anus. Its wall contains an inner sphincter of plain muscle and an outer sphincter of striped muscle, which control the opening and closing of the anus.

Movements of the Large Intestine

The intestinal contents passively fill the caecum and

ascending colon, and further movement along the colon is achieved mainly by series of muscular relaxations of the intestinal wall.

Functions of the Large Intestine

The large intestine absorbs water, sodium and chloride from the intestinal contents, and excretes potassium into them. The amount of water absorbed is up to 2 litres in 24 hours, and as a result of this absorption the faeces become solid.

Defaecation

Defaecation is partly a reflex action, partly a voluntary one. A desire to defaecate is stimulated by the entry of faeces into the rectum. In the act of defaecation; the anal sphincters relax, the muscle in the wall of the large intestine contracts, the pelvic floor is pulled up by muscular action and finally the anal sphincters contract, expelling the last of the faeces.

Faeces

About 100g of faeces are passed daily. They are composed of water, cellulose, calcium, fat, inorganic salts, derivatives of bile pigments, mucus and micro-organisms.

THE PERITONEUM

The peritoneum is a thin membrane, enclosing with one layer many of the abdominal organs and with another lining the walls of the abdominal cavity. A film of fluid enables one layer to glide smoothly over the other. The peritoneal cavity is the potential space between the two layers, i.e. there is normally no space as such, but one could be filled with air or additional fluid if they got between the two layers.

The arrangement of the peritoneum is complex (see Fig. 10/5). The *greater sac* forms the greater part of it and communicates with the *smaller sac*, which is behind the stomach. The stomach is enclosed within two layers. The *mesentery* is a double layer of peritoneum connecting the jejunum and ileum to the back of the abdomen; it is short behind and fans out to about 6 metres wide at its attachment to the intestine and has to be packed into many folds. The large intestine and the liver are largely enclosed within

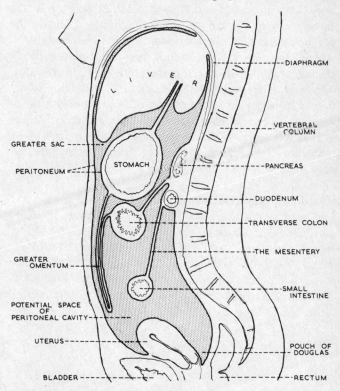

Fig. 10/5 The peritoneum, in the female body

peritoneum. The *great omentum* is a large four-layer flap of peritoneum which hangs down like an apron from the stomach and transverse colon; in fat people it becomes a large deposit of fat. Many of the abdominal organs are attached to each other or to the abdominal wall by folds of peritoneum. The kidneys and other structures at the back of the abdomen lie behind the peritoneum.

In the *male pelvis* the peritoneum passes forwards from the rectum on to the bladder, forming a pouch called the recto-vesical pouch, and covers the sides and top of the bladder, being pushed up by the bladder as it fills. In the *female pelvis* the peritoneum passes off the rectum on to the uterus, forming a pouch called the recto-uterine pouch (pouch of Douglas), and from the uterus on to the bladder. The *broad ligament* is a double layer of peritoneum which passes from the uterus to the pelvic wall on each side; it encloses between its layers the blood-vessels, lymph-vessels and nerves of the uterus, and the uterine (Fallopian) tubes, the end of each tube having a tiny opening into the peritoneal cavity.

Functions of the Peritoneum

The peritoneum is a tissue of great importance. Its functions are (a) to maintain the organs in their places; (b) to enable the stomach and intestines to move; (c) to provide the means by which blood-vessels, lymph-vessels and nerves can reach organs without being kinked or obstructed, and (d) to check and seal off any infection within the abdomen.

Chapter 11

NUTRITION AND
METABOLISM

Food has to provide energy for the body, to build and maintain the tissues, and to provide the chemical substances necessary for bodily processes. A diet must contain adequate amounts of protein, fat, carbohydrate, vitamins, minerals and water.

1. Protein
Proteins are necessary for growth and replacement of worn-out and damaged tissues.

The principal sources of protein are meat, fish, eggs, peas and beans.

A protein is composed of amino-acids. Some proteins can be manufactured in the body as well as being present in food. Enzymes in the alimentary tract break proteins down into amino-acids, which are absorbed through the cells of the wall of the small intestine. The cells of tissues which require certain amino-acids have the ability to abstract them from the blood and to build them up into the necessary proteins. Any excess of amino-acid is broken down by the liver and excreted as urea by the kidneys.

2. Fat
Fats are necessary as a souce of energy, a protection for some of the organs of the body, and a source of the fat-soluble vitamins A, D, and K.

Principal sources are milk, cream, butter, egg-yolk,

animal fats, vegetable oils. Margarine is a synthetic fat to which vitamins A and D have been added.

Fats are broken down into smaller droplets and fatty acids by the action of bile salts and lipase, an enzyme in pancreatic juice. Fatty acids are absorbed from the small intestine into the portal veins, but some fat particles pass into the lymph-vessels of the small intestine and reach the blood by passing first along the thoracic duct.

Fats can be broken down to produce energy, used in cells, stored in fat depots of the body, of which the main ones are in the abdomen and under the skin, and converted into glucose.

When fat is required, it can be withdrawn from the fat depots and converted by the liver into fatty acids and glycerol.

3. Carbohydrate

Carbohydrates are a source of energy. They are composed of carbon, hydrogen and oxygen, and occur as sugars and starches.

As *sugar* they are present in cane and beet sugar, honey, jam and dried fruits, and as *starch* in bread, cake, cereals and potatoes.

They occur in three main types:

1. Monosaccharides (one-sugar groups), e.g. glucose, fructose.
2. Disaccharides (two-sugar groups), e.g. sucrose (cane and beet sugar), lactose (the sugar in milk).
3. Polysaccharides (multiple-sugar groups), e.g. starch (in plants), glycogen (in animal tissues, especially liver and muscle).

Cellulose, a polysaccharide, forms the fibre of plants; it cannot be digested by man and passes unchanged through his alimentary tract.

Carbohydrates can pass through the wall of the small

intestine as glucose, fructose and galactose, and other types have to be converted into these by enzyme action. In the body fructose and galactose are converted into glucose, which becomes the final stage of all carbohydrates. Glucose passes from the liver into the blood, and after a meal there is a rise for a time in the amount of glucose in the blood. The glucose in the blood is commonly called the blood sugar.

Glucose can be used as a source of energy, converted into glycogen, which is stored in the liver and muscles, or converted into fat.

Glycogen in the liver is changed back to glucose when there is a need for more glucose, and glycogen in muscle can be used as a source of energy for muscular contractions.

4. Vitamins

The vitamins are chemical substances which in small amounts are necessary for the health of the body and for particular functions. They are not a source of energy. They are divided into (a) fat soluble vitamins: A, D, K and (b) water soluble vitamins: B complex, C.

1. Vitamin A
Sources: animal fats, fish liver oils.
Functions: it is necessary for the health of epithelium and for the regeneration of visual purple in the retina of the eye after it has been bleached by light.

2. Vitamin D
Sources: animal fats, fish liver oils, flesh of fatty fish (identifiable by being pinky-brown).
Functions: it is necessary for normal bone growth; lack of it in childhood causes rickets.

3. Vitamin K
Sources: meat, green vegetables.
Function: it is necessary for the clotting of blood.

4. Vitamin B complex

The original vitamin B was found to be composed of several vitamins, including thiamine, nicotinic acid, riboflavin, folic acid and vitamin B_{12}.

Sources: liver, kidney, eggs, yeast, beans, nuts, etc.

Functions:

> *Thiamine*: lack of it causes beri-beri, a disease in which the brain, nerves and heart are affected.
>
> *Nicotinic acid*: lack of it causes pellagra, a disease in which the skin, intestine and brain are affected.
>
> *Riboflavin*: it is necessary for the health of epithelium.
>
> *Folic acid*: it is necessary for red blood formation.
>
> *Vitamin B_{12}*: it is necessary for red blood formation.

5. Vitamin C (ascorbic acid)

Sources: fresh fruit and vegetables, especially oranges, tomatoes and grapefruit.

Functions: it is necessary for the health of the capillary wall. Lack of it causes scurvy, a disease characterised by haemorrhages.

5. Minerals

The body needs many minerals, e.g. calcium, phosphorus, sodium, potassium, iron, copper, cobalt, iodine, magnesium. They must be present in the diet both in the right amounts and in forms that can be absorbed and utilised. Some, such as calcium and phosphorus, are needed in relatively large amounts, some, such as copper and cobalt, only in traces. They are not usually absent from a normal diet as most of them are present in many kinds of food. There is an increased demand for them during growth, pregnancy and lactation.

Calcium is present in milk, cheese and butter. Its absorption in the small intestine is dependent upon the presence of adequate amounts of phosphorus and vitamin D. There is a marked reduction of absorption in old age. It is necessary for (a) the development of bone and teeth, (b) the clotting of blood, (c) the contraction of heart muscle, and (d) the transmission of impulses from nerve to muscle at the neuro-muscular junction.

Phosphorus is present in milk, eggs and liver. It is necessary for the maintenance of the normal acid-base balance of the body, the provision of energy for muscular contraction and the development of bone.

Potassium is present in most foods, especially fresh orange juice. It is necessary for the activity of the cell membrane and the passage of fluid through it, and the contraction of muscle.

Iron is present principally in beef, liver, spinach. It is necessary for the development of red blood cells.

Only about 10 per cent of the iron in an average diet is absorbed. The body utilises any iron left over when old red blood cells are destroyed and uses it to form new haemo globin in new red blood cells. A little of the old iron is excreted, and it is the loss of this amount that has to be made up from food. Women absorb more iron than men because they need to replace that lost in menstruation.

Copper and *Cobalt* are necessary in trace amounts for the formation of red blood cells.

Iodine is present in sea-food and crops grown near the sea. It is necessary for the formation of two hormones of the thyroid gland.

Magnesium is necessary in trace amounts for muscular activity.

6. *Water*

Water is vital. Without it life cannot be sustained for more than a few days. The body of an average man weighing 70 kilos contains about 50 litres of water. It is present in cells, in the extracellular fluid around cells, in the plasma of the blood, and in glandular secretions. Its functions include (a) holding many substances in solution, (b) being a carrier in emulsion of substances it cannot dissolve, and (c) having a high capacity for absorbing heat, so that a lot of heat can be produced with little alteration in the temperature of the body.

An adult engaged in a sedentary occupation in a temperate climate requires about 2.5 litres a day. He gets it from food and drink and by the oxidation of food in the body. He gets it in about the following amounts:

Drink	1 300ml
Food	850ml
Oxidation	350ml

In hot countries and when he is doing hard work he will need to drink more to replace the greater loss of fluid.

A BALANCED DIET

A balanced diet is one that contains the various food-stuffs in appropriate amounts and proportions for normal growth, the repair of tissues, the maintenance of an appropriate weight and the maintenance of health. The amounts and kinds of food eaten vary with the food available, local custom, and a person's income, age, knowledge of food values, health, intelligence and emotional balance. The requirements are measured in Calories. (A Calorie (big C) = 1 000 calories (small c). 1 c is the amount of heat required to raise 1 gram of water by one degree Celsius.) Typical requirements are:

man doing light work	2 400 C per day
man doing medium work	3 300 C per day
man doing heavy work	4 500 C per day
woman doing light work	2 100 C per day.

Proteins are provided mainly by meat, fish, eggs, poultry,

milk, cheese and nuts; fats by butter, margarine, cream, oils, eggs and fat meat; carbohydrates by sugar, potato, bread, cereals and other grain products. Adequate amounts of these will provide all the minerals and vitamins with the possible exception of vitamin C, which is quickly destroyed by cooking and to provide which fresh fruit or vegetables have to be taken.

In affluent societies people tend to eat more protein than they need as well as too much food altogether. An excess of carbohydrate is bad as the sole function of carbohydrate is to provide energy, which can be adequately provided by protein and fat. Any excess of carbohydrate is laid down as fat. Moreover, many carbohydrate foods have been 'refined', i.e. had much of the fibre removed that is normally present in the outer layers of seeds, as well as in fruit and vegetables. Fibre is necessary to provide an adequate bulk for the stools, to render the work of the colon more efficient, and to make defaecation easy. The diet should therefore include wholemeal bread, bran, fruit and vegetables.

MILK

The young mammal begins life by drinking milk whose composition is appropriate to its rate of growth. This can be seen in a comparison of human and cow's milk, which have the following average composition:

	Human Milk per cent	*Cow's Milk* per cent
Protein	1.7	3.2
Lactose	7.5	4.8
Milk fat	3.5	3.7
Mineral salts	0.2	0.7
Water	87.1	87.6

Because a calf grows much more quickly than a baby, cow's milk contains about twice as much protein and about four times as much calcium as human milk. For normal growth in infancy human milk contains adequate amounts of protein, fat, carbohydrate, vitamins and minerals. It contains only a little iron, but a baby is born with a store of iron big enough to provide what it needs in the first six months of life; over that age a baby fed only on milk is liable to become anaemic. Cow's milk has the disadvantages for a baby of having too much protein, too little carbohydrate and too little vitamin C.

Chapter 12

THE LIVER AND THE
PANCREAS

THE LIVER

The liver, the largest gland in the body, lies in the upper right region of the abdomen just below the diaphragm, being protected in front and at the side by the lower ribs and costal cartilages. It projects slightly beyond the right costal margin, but it is too soft to be felt. It is wedge-shaped with its base to the right and its apex tapering over the midline to the left. It is almost entirely covered with peritoneum, folds of which connect it to the stomach and other structures. It is imperfectly divided into a large right lobe and a small left lobe. The gall-bladder is attached to its inferior surface and projects slightly beyond its anterior border (see Fig. 12/2). The inferior vena cava runs in a groove at the back of the liver.

The liver has a double blood supply: (a) the hepatic artery which brings oxygenated blood to it from a branch of the aorta, and (b) the portal vein which brings venous blood from the stomach, intestines and spleen.

Within the liver the streams of arterial blood and venous blood mix and run through sinusoids, the blood being thus in direct contact with the liver cells. Having passed through the sinusoids, the blood returns to the general circulation by passing through two hepatic veins into the inferior vena cava as it passes upwards at the back of the liver.

The liver is composed of a large number of small lobules,

each composed of hepatic cells arranged in columns. The blood runs in sinusoids between these columns. Between adjacent columns of cells run the small canals which form part of the biliary system.

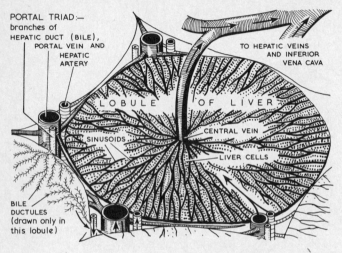

PORTAL TRIAD:—
branches of
HEPATIC DUCT (BILE),
PORTAL VEIN AND
HEPATIC
ARTERY

TO HEPATIC VEINS
AND INFERIOR
VENA CAVA

LOBULE OF LIVER

SINUSOIDS

CENTRAL VEIN

LIVER CELLS

BILE
DUCTULES
(drawn only in
this lobule)

Fig. 12/1 The liver cells and sinusoids

The Biliary System

The biliary system is the system through which bile passes on its way out of the liver (see Fig. 12/2). The small canals into which the bile passes from the liver cells join to make larger canals and these join to form a right and left hepatic duct. These two join to form a common hepatic duct, a tube about 3cm long. The common hepatic duct is connected by another tube, the cystic duct, to the gall-bladder.

The *gall-bladder* is a sac about the size and shape of a small pear. Its end is rounded and projects beyond the liver just below the right costal margin; its body is attached to the inferior surface of the liver, and its narrow neck is continuous with the cystic duct. It can hold about 50ml of bile.

The *bile duct*, formed by the junction of the hepatic and cystic ducts, runs downwards behind the first part of the duodenum and then through the head of the pancreas to pass with the pancreatic duct through the wall of the descending part of the duodenum. There the two ducts unite in a small dilatation called the ampulla of Vater, which opens into the interior of the duodenum, the opening being controlled by a little sphincter of muscle.

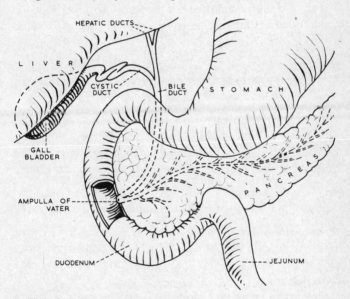

Fig. 12/2 The biliary system, pancreas and pancreatic duct

The Functions of the Liver
1. Carbohydrate Metabolism
Glucose is stored in the liver in the form of glycogen. Glycogen is a carbohydrate composed of many hundreds of

units of glucose. Glucose is converted into glycogen by the action of insulin. Glycogen is converted into glucose by the action of adrenaline and glucagon.

2. Fat Metabolism
Fat is broken down in the liver into fatty acids and glycerol, and fatty acids are converted into types which can be used in metabolic processes.

3. Protein Metabolism
Some amino-acids are converted in the liver into glucose. Unwanted amino-acids are converted into urea and uric acid, and these are discharged into the blood and excreted by the kidneys.

4. Storage
The liver is a store of glycogen, fat, the fat soluble vitamins A, D and K, vitamin B_{12} and iron.

5. Red blood cells
The liver produces red blood cells in fetal life only. Throughout life red blood cells are destroyed in the reticulo-endothelial cells, some of which line the sinusoids.

6. Plasma proteins
Some of the plasma proteins are manufactured in the liver.

7. Detoxication
The liver breaks down steroid hormones and many drugs.

The Bile
The bile is an alkaline, sticky fluid, with a yellow or green colour and a bitter taste. It is secreted continuously by the liver cells at a rate of 500–1000ml a day. It passes through the hepatic and cystic ducts into the gall-bladder, where it is concentrated. The arrival of fatty food in the duodenum stimulates the secretion of a hormone called cholecystokinin from duodenal cells and this hormone, circulating in the

blood, causes the gall-bladder to contract and drive its contents into the duodenum.

Bile is composed of water, bile salts, bile pigments, cholesterol and inorganic salts. The *bile salts* have an important role to play in digestion. They alter the tension on the surface of fat globules, making them smaller and exposing them to enzyme action, they stimulate muscle movements in the intestine, and after being reabsorbed from the intestine they stimulate the formation of more bile. The *bile pigments* are mainly derived from the break-down of haemoglobin. They have no digestive functions. Some of the pigment is converted in the intestine into stercobilin, which gives faeces their brown colour.

THE PANCREAS

The pancreas is a long, flat gland which lies across the back of the abdomen. It has a large, rounded head lying in the curve of the duodenum, a long body behind the stomach, and a tail which touches the spleen (see Fig. 10/4). The pancreatic duct runs the length of the gland, receiving smaller ducts from the lobules of the gland, and is joined by the bile duct, the two ducts running through the head of the pancreas and opening by a common opening into the duodenum (see Fig. 12/2).

The pancreas is composed of two organs in one, but this cannot be appreciated by the naked eye. It is composed of:
1. Cells which produce pancreatic juice. These cells are grouped in lobules, and the juice they produce passes into the pancreatic duct and so into the duodenum. *Pancreatic juice* is a clear alkaline fluid containing enzymes which act on fat, protein and carbohydrate. The arrival of the gastric contents in the duodenum is the signal to the pancreas to start secreting its juice. When the pancreatic juice mixes with the partly digested food

in the small intestine, its enzymes break proteins down into amino-acids, sugars into glucose, and fats into fatty acids and glycerol. It neutralises the hydrochloric acid of gastric juice.

2. The islets of Langerhans. These are clumps of cells lying between the lobules and having no connection with the cells that secrete pancreatic juice. They are composed of two types of cell: alpha cells which secrete glucagon, and beta cells which secrete insulin (see chapter 18).

Chapter 13

THE URINARY SYSTEM: WATER BALANCE

The urinary system consists of the kidneys, the ureters, the bladder and the urethra (see Fig. 13/1).

THE KIDNEYS

The kidneys, right and left, lie at the back of the abdomen, behind the peritoneum and on either side of the vertebral column. Each is about 11cm long, 6cm broad and 3cm thick. They are embedded in fat, which separates them from the abdominal organs in front and the posterior abdominal wall behind, and protects them from injury. An adrenal gland is perched on the top of each kidney (see Fig. 10/4, p. 106). The concave inner border is marked by the hilum, a deep fissure through which pass the renal blood vessels and nerves and to which is attached the pelvis of the ureter. A wide renal artery passes from the aorta to each kidney, and a wide renal vein from each kidney to the inferior vena cava; they are wide because at each beat of the heart about a quarter of the blood has to enter the kidneys. The surface of the kidney is enclosed in a fibrous capsule. When a kidney is cut across it is seen to consist of a cortex, an outer darker part, and a medulla, an inner part of cones of renal tissue projecting into the hilum.

The Nephrons

A *nephron* is the functional unit of the kidney. Each kidney

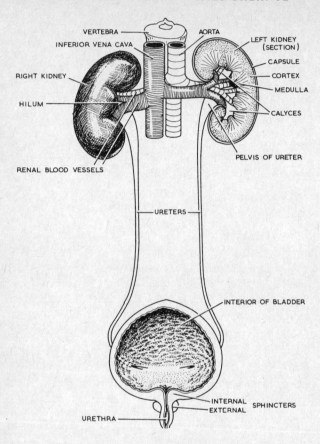

Fig. 13/1 The urinary system

is composed of about one million of them. A nephron is a
long, twisted tube, one end of which is blind and the other
opening with other nephrons into a larger collecting tubule
(see Fig. 13/2). The blind end, called Bowman's capsule, is
like a thin-walled bag punched in until the surfaces nearly
touch, and contains in the cavity the *glomerulus*, a whorl of

capillaries. From Bowman's capsule the nephron is continued as the first convoluted tubule, then as the first loop of Henle, which runs into and out of the medulla, then as the second convoluted tubule, and finally as a tube running into the collecting tubule. The collecting tubules open on the surface of the cones of renal tissue within the pelvis of the ureter.

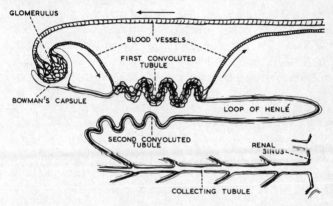

Fig. 13/2 A nephron

The blood supply to the nephrons is distinctive. Arterial blood runs into the capillaries which form the glomerulus of each nephron, but when it leaves the glomerulus it does not pass immediately into a venule but forms a network of capillaries over the renal tubule of the same nephron, and only then does it pass into a venule. It is important to appreciate that the blood that has passed through a glomerulus then passes over the tubule.

THE URETERS

The ureters are two tubes, right and left, which connect the kidneys to the bladder. Each is about 25–30cm long, begins as the pelvis of the ureter (a funnel–shaped enlargement attached to the edge of the hilum of the kidney) runs down the back of the abdomen, passes into the pelvis, and runs obliquely through the wall of the bladder to open on its interior surface.

THE BLADDER

The bladder, the receptacle into which the urine passes from the ureters, lies when empty in the pelvis. In front it has the pubis. In the male it is in front of the rectum (see Fig. 19/1, p. 187), in the female it is in front of the vagina, with the uterus lying partly above and partly behind (see Fig. 19/2, p. 189). When more than half distended with urine it rises up into the abdomen above the pubis, pushing up the peritoneum as it rises.

The interior of the bladder is the same in both sexes. It is lined with pink mucous membrane, loosely connected with a muscular coat and thrown into folds when the bladder is empty. There are three openings into the bladder: (a) the orifices of the ureters, and (b) the internal orifice of the urethra. All these orifices are about 2.5cm apart in the empty bladder.

THE URETHRA

The urethra is the tube from the bladder to the surface of the body.

The *male urethra* is about 20cm long and extends from the internal urethral orifice at the bladder to the external urethral orifice at the end of the penis. Just below the bladder it is surrounded by the prostate gland, which in late adult life is liable to become enlarged and prevent urine from leaving the bladder. The passing of urine through the urethra is controlled by the contraction and relaxation of two sphincters – an internal sphincter of involuntary muscle which surrounds it as it leaves the bladder and an external sphincter of voluntary muscle fibres just below the prostate gland.

The *female urethra* is about 4cm long. It begins at the internal urethral orifice in the bladder, runs downwards in front of the vagina, and opens at the vestibule immediately in front of the vagina. Like the male urethra it has internal and external sphincters.

THE FUNCTIONS OF THE KIDNEYS

The kidneys are the principal excretory organs of the body. Their functions are:

1. To maintain within normal limits the acid-base balance of the body and the chemical composition of the fluids of the body and in particular the amounts of sodium and potassium in them.
2. To excrete some of the end-products of metabolism: *urea*, produced by the breakdown of amino-acids, either from food or from cells; *creatinine*, produced by the breakdown of muscle fibres; and *uric acid*, derived from the nucleic acid of food or broken down cells.
3. In addition, the kidneys excrete some drugs, play a role in the production of vitamin D, and produce an enzyme which leads to a rise of blood pressure and a hormone which stimulates the production of red blood cells.

The Secretion of Urine

Urine is produced from the blood by the action of cells of the nephron. Production is carried out in two stages, and to understand it we must remember the anatomy of a nephron and the way in which the blood from a glomerulus passes into a plexus of veins over the tubules before it leaves the kidney.

1. The *first stage* is the filtering of a fluid out of the blood in the glomerulus into the space between the two layers of Bowman's capsule. This fluid is different from urine in several ways: it is very dilute and contains, among other things, glucose, which normal urine does not contain.

2. The *second stage* takes place in the tubule. In the loop of Henle some of the salts, most of the water and all the glucose are reabsorbed into the blood, just as if the body had had second thoughts and realised it was going to lose some of the things it really required. The amount of water absorbed (it can be as much as 147 litres in 24 hours) is partly controlled by an anti-diuretic hormone produced by the hypothalamus-pituitary complex (diuresis is the production of excessive amounts of urine). As the fluid passes through the second convo-luted tubule some other substances are added to it, and the fluid that now passes along the collecting tubules into the pelvis of the ureter is urine.

The Urine

The volume and composition of the urine vary from person to person and from day to day in the same person. These variations are the result of changes in the amount of fluid drunk, the diet, the temperature and humidity of the air, the amount of fluid lost through the skin, lungs and alimen-tary tract, and the amount of muscular exercise taken. More is secreted by day than by night.

1. *Amount*: 900–1 500ml per 24 hours.
2. *Reaction*: acid, when a person takes an ordinary diet.
3. *Specific gravity* (SG): 1 002–1 032, varying with the amount of solids in it.
4. *Composition*: water, urea (20–30g in 24 hours), uric acid, creatinine, ammonia, sodium, potassium, chloride, phosphates, sulphates.
5. *Colour*: due to a pigment called urochrome.

Micturition

Having been passed down the ureters by peristaltic waves, urine accumulates in the bladder. A relaxation of the tone of the muscle of the bladder wall allows the bladder to enlarge considerably before the pressure in it becomes raised, but an individual limit is reached at which the pressure rises, sensory nerves in the bladder wall are stimulated, and a desire to micturate is felt. At this stage rhythmic contractions begin in the muscle.

The act of micturition is essentially a reflex action, which after infancy is controlled by centres higher in the nervous system. In it the muscle of the bladder wall contracts and the sphincters relax. The expulsion of urine is assisted by contraction of the muscles of the abdominal wall and of the diaphragm. The last drops of urine are squeezed out of the urethra by the contraction of a small muscle just below the bladder.

WATER BALANCE

The water balance of the body must be maintained by equality of intake and output. As water is being lost continuously through the kidneys, the skin and the lungs, the main problem is to keep enough water in the body.

The External Exchange of Water

In a temperate climate a person's intake of water is about 2.5 litres a day, obtained from:

Drink	1	300ml
Food		850ml
Oxidation of food		350ml

His excretion is approximately:

In the urine	1 500ml
Through the skin	500ml
By the lungs	400ml
In faeces	200ml

The Internal Exchange of Water

Important exchanges of water are continuously taking place in the body:

1. *In the alimentary tract*: water is taken out of the blood to make saliva, gastric juice, bile, pancreatic juice, and intestinal juice. Most of this water is reabsorbed from the small and large intestine and returned to the blood.
2. *In the tissue spaces*, there is an exchange of water in both directions between the tissue spaces and the capillaries.
3. *In the brain*, cerebro-spinal fluid is formed out of the blood in the choroid plexuses and after circulating round the brain and spinal cord is returned to the blood.
4. *In the kidneys*, most of the water filtered out of the blood in the glomeruli is reabsorbed as it passes through the renal tubules.

Osmotic Pressure

Osmotic pressure, the pressure under which fluids pass through a membrane in proportion to the concentration of molecules in them, is important in life because cell walls act as membranes. The osmotic pressure of the plasma must be kept constant, which is achieved mainly by the action of the plasma proteins. If the plasma was to become too strong, fluid would be drawn out of the blood cells and tissue spaces, and if it was to become too weak, fluid would pass out of it into the cells, the blood cells swelling and possibly bursting.

An isotonic solution is one of the same osmotic pressure

as plasma. A hypertonic solution is one whose osmotic pressure is higher than that of plasma, and a hypotonic solution is one whose osmotic pressure is less than that of plasma.

Chapter 14

THE SKIN: THE TEMPERATURE OF THE BODY

THE SKIN
The skin consists of two layers: the epidermis and the dermis (see Fig. 14/1).

The Epidermis
The epidermis is the outer layer of skin. It is composed of cells which become flattened as they reach the surface, where they are worn away by friction. It has no blood vessels and so does not bleed when cut. Its thickness is variable: on the eyelids it is very thin, on the palms and soles very thick. The parts exposed to heavy wear have a high content of keratin, a horny protein. The surface of the epidermis is marked by a number of narrow ridges arranged in groups of parallel lines in loops and other patterns. They are constant throughout life and are the basis of identification by finger, palm or sole prints.

The colour of a person's skin is due to the presence of granules of melanin, a black pigment, in the deeper layers of the epidermis. The more pigment the darker the skin.

The Dermis
The dermis is thicker than the epidermis. Like the epidermis it is thicker in areas exposed to friction and pressure. Its outer layer projects in ridges and papillae into the

epidermis; its inner surface is in contact with the subcutaneous tissue. It is composed of connective tissue and
contains many blood-vessels, lymph-vessels and nerves.

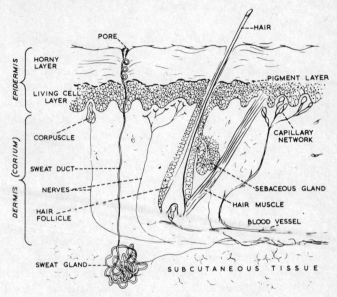

Fig. 14/1 A section through the skin

The *nails* are pieces of horny tissue, firmly attached to
the dermis at the nail-bed. The little 'moon' at the growing
end is denser than the rest.

The *hairs* are found in most parts of the skin, but they are
not present in the palms and soles. They vary in length,
thickness and colour. A hair grows from a hair-follicle in
the dermis. A hair-follicle is a shaft at the bottom of which
the hair grows out of a small projection. A few involuntary
muscle-fibres are attached to the exterior of the follicle and
can by contracting make the hair stand up; cold and fear
can stimulate these muscles and so cause 'goose flesh'. A

sebaceous gland is a small gland attached to a hair-follicle into which it opens by a small duct. It secretes sebum, a fatty substance which coats each hair. Sebaceous glands are particularly numerous in the scalp and other hairy regions.

A *sweat-gland* is a single coiled tube situated in the subcutaneous tissue and with a long duct which runs through the skin to open on its surface. There are two kinds of sweat-glands – ordinary sweat-glands and the apocrine glands in the axilla and vulva, which produce a secretion that develops a particular smell as a result of bacterial action. The glands that produce wax in the ears are modified sweat-glands.

Nerve endings in the skin

The skin is supplied with sensory and sympathetic nerves. The sensory nerve-endings take the form of (a) free nerve-endings, with no particular structure at the tip, (b) nerve-endings around follicles, (c) Meissner's corpuscles, small rounded structures around nerve-endings in the papillae of the dermis, and (d) Pacinian corpuscles, large encapsulated structures in the deeper layers. The sympathetic nerves supply the arterioles, the sweat glands, and the muscles attached to the hair follicles.

The Functions of the Skin

1. *Protection* The skin protects the internal organs against injury and invasion by harmful micro-organisms, which have difficulty in getting through it.
2. *Sensation* Sensations of touch, pain, changes of temperature and pressure are appreciated in the skin and subcutaneous tissue and transmitted through sensory nerves to the spinal cord and brain.

   The functions of the various nerve-endings in the skin and subcutaneous tissue are not definitely known. In any area of skin at any one time there are tiny spots in which touch, pain and temperature are individually appreciated, but these spots change from day to day – a spot which appreciates temperature one day may be found to be appreciating touch another. The Pacinian corpuscles may be the organs in which pressure changes are appreciated. Not all nerve-endings may be functioning at any one time; possibly they have rest periods.

3. *Heat regulation*   (see below).
4. *Storage*   The skin has a store of fat and water, which can be drawn on when required.
5. *Absorption*   The skin absorbs ultra-violet light, which stimulates the production of vitamin D in the skin.

THE TEMPERATURE OF THE BODY

For temperature purposes the body can be considered as composed of (a) an inner core – the inside of the chest and abdomen, and (b) a peripheral shell – the limbs, skin, subcutaneous tissue and muscles.

The temperature of the inner core is kept fairly constant, but the temperature of the peripheral shell can vary.

The normal peripheral temperature of the body varies between 36° and 37.5°C. The figure of 37°C marked on thermometers and temperature charts is an average and not a constant figure. There is a daily variation in temperature, which is at its highest in the afternoon and its lowest between 02.00 and 06.00 hours. An individual has a steady pattern of temperature variation, which does not vary with the seasons nor with working at night. In women there is also a monthly variation, the temperature during the first half of the menstrual cycle being lower than that during the second half. Temperature taken in the rectum is higher than that taken in the mouth or axilla and closer to the core temperature, which is about 37.8°C.

The body keeps its temperature within the normal range by balancing its heat gain and heat loss.

Heat Gain

Heat is gained by being produced in the body and taken up from the environment.

1. *Heat production* All the metabolic activities of the body produce heat. That produced by the muscles is increased by exercise. The amount produced by the other organs is fairly constant. There is a relation between heat production and the surface area of the body; the greater the relative surface area – as in an infant – the greater must be the heat production to compensate for the greater heat loss.

2. *Heat from the environment* The body takes in heat from the sun, from radiation from the sky, from hot food and drinks, from hot baths, from hot air in hot climates, and from hot soil when it is in contact with the body.

Heat Loss

Heat is lost from the skin, in expired air, and in the urine and faeces.

1. *Sweat* The greater the sweat production the greater the loss of heat, provided that the sweat is not wiped off, for heat is lost only if it is allowed to evaporate.

2. *Insensible perspiration* Heat is lost with the water that diffuses through the skin. Insensible perspiration takes place continuously and is little affected by the environment. It is called insensible because it is neither felt nor seen.

3. *Other losses* Heat is lost in expired air, urine and faeces.

Chapter 15

THE LYMPHATIC SYSTEM

The lymphatic system is composed of lymph-capillaries, lymph-vessels, lymph-nodes, other lymph-tissue and lymph.

Lymph-Capillaries

Lymph-capillaries are thin-walled tubes lying in the tissue spaces of various organs and tissues. There are none in the central nervous system. The walls are thin for the passage through them of fluid and particles too big to pass through the wall of a capillary.

Lymph-Vessels

Lymph-vessels are formed by the union of lymph-capillaries. They run into lymph-nodes and contain valves which give them a beaded appearance.

Lymph-Nodes

Lymph-nodes are small round or oval masses of lymph-tissue enclosed in a capsule (see Fig. 15/1). They are arranged in groups and chains. Each receives several lymph-vessels, which transmit lymph to them. From them the lymph is transmitted through another lymph-vessel to another lymph-node.

The lymph-tissue of which a node is composed consists of lymph-cells, from which are produced the lymphocytes of the blood.

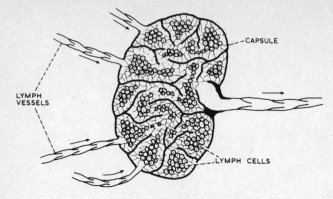

Fig. 15/1 A lymph gland and vessels

The largest groups of nodes are:
(a) In the neck, draining the head and neck
(b) In the axilla, draining the arm and breast
(c) Around the roots of the lungs, draining the lungs and heart
(d) In the abdomen, and especially in front of the aorta and along its branches, draining the abdominal organs
(e) In the pelvis, draining the pelvic organs
(f) Behind the knee and in the groin, draining the leg.

Lacteals
The lacteals are the lymph-vessels which drain the small intestine. They become milky-white with fat globules during the absorption of fat from the intestine, and get their name from the Latin word for milk.

Receptaculum Chyli and Thoracic Duct
The receptaculum chyli is a sac at the back of the abdomen. It receives lymph from the abdominal and pelvic nodes. The thoracic duct is a continuation upwards from the receptaculum chyli. In a long course it passes through the

diaphragm and chest to end in the neck by joining the junction of the left subclavian and left internal jugular veins, where the lymph it conveys is discharged into the blood (see Fig. 15/2).

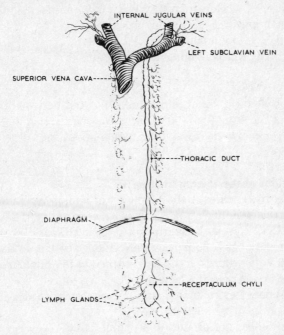

Fig. 15/2 The thoracic duct

Other Lymph-Tissue

This tissue is identical with that of the lymph-nodes and occurs in several parts of the body:

(a) The *tonsils* are two oval patches of lymph-tissue situated between the anterior and posterior pillars of the fauces, which are two folds of mucous membrane on either side of the entrance into the pharynx from the mouth.

(b) The *adenoids* are patches of lymph-tissue, variable in size, in the nasopharynx. In children they are liable to become enlarged and to interfere with breathing through the nose.

(c) *Peyer's patches* are oval patches of lymph-tissue in the small intestine, especially the ileum.

(d) The *spleen* contains much lymph-tissue.

(e) The *thymus* contains much lymph-tissue.

Lymph

Lymph is a transparent, slightly alkaline fluid, derived from plasma and from the fluid of the tissue spaces. It contains lymphocytes on their way from lymph-nodes and tissue to the blood.

Functions of the Lymphatic System

1. The tissue spaces are drained into the lymph-capillaries. Some of the fluid is reabsorbed by the blood, but the rest and particles (such as carbon inspired in air) pass into the lymph-nodes. The particles are trapped in the nodes.

2. About two-thirds of the fat absorbed in the small intestine passes into the lacteals.

3. Lymphocytes are manufactured in the lymph-nodes and tissue.

4. Antibodies are produced in lymph-tissue.

Chapter 16

THE NERVOUS SYSTEM

The nervous system is essentially an organ for doing three things:
1. Receiving sensations from both the external world and the inside of the body.
2. Storing them, thinking about them, associating them with other impressions, and remembering them.
3. Acting upon them by sending messages to muscles, glands and other organs.

The Central and Peripheral Nervous System

The *central nervous system* consists of the brain and spinal cord. The *peripheral nervous system* consists of 12 pairs of cranial nerves attached to the brain and 31 pairs of spinal nerves attached to the spinal cord.

Sensory (afferent) nerves are the nerves that carry messages to the brain and spinal cord. These messages come through any of the special senses (feeling, seeing, hearing, tasting, smelling) or from structures inside the body, such as the muscles, tendons and glands.

Motor (efferent) nerves are the nerves that convey messages from the brain and spinal cord to the muscles, glands, and other organs, stimulating them to carry out their various functions.

The Autonomic Nervous System

The autonomic nervous system consists of two parts – the sympathetic and the parasympathetic systems. It is the part of the nervous system concerned with the unconscious performance of internal organs, such as the heart, the blood-vessels, the lungs, the alimentary tract.

NERVOUS TISSUE

Nervous tissue is composed of *neurones* which are nerve-cells and their attached fibres and *neuroglia* which are cells whose functions are not definitely known; they may be involved in absorbing micro-organisms and other foreign substances should they get into nervous tissue.

Neurone

A *neurone* consists of a nerve-cell and its attached fibres, and is the basic unit of all nervous tissue (see Fig. 16/1).

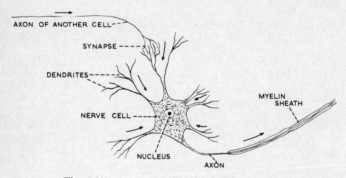

Fig. 16/1 A nerve cell, showing a synapse

Nerve-cells vary in size and appearance, according to their various functions. They form the grey matter of the nervous system. A nerve-cell has a well marked nucleus, and its protoplasm may show masses of granules and some fine fibres called fibrils. The *dendrites* are branched processes which project from the surface of the cell and are in contact with other nerve cells and their dendrites; they are capable of receiving nervous impulses and of passing them into the cell, but not of transmitting them in the opposite direction. The *axon* is a single process through which impulses are transmitted out of a cell. Some axons can be long, e.g. those passing from the top of the brain to the spinal cord, or from the spinal cord to the hand or foot. These axons are the nerve-fibres which form the white matter of the nervous system. Either before birth or within a few months of birth, most fibres become myelinated, i.e. enclosed within a sheath of myelin, a fatty substance; it is thought that a fibre cannot function properly until it is myelinated.

Nerve-cells cannot reproduce themselves, and the number of cells in the body is fixed before birth. If one dies, it cannot be replaced, but in some circumstances other nerve cells may take over its functions.

Each neurone is structurally an independent unit, but neurones are functionally dependent upon one another in the transmission of nervous impulses.

A *nerve-impulse* is a complex chemico-electrical charge which travels along a nerve fibre. A *synapse* is the point of communication of one neurone with another, and at this point the further transmission is done chemically, a chemical change being produced which transmits the impulse to the next neurone.

THE CENTRAL NERVOUS SYSTEM

The central nervous system is composed of the brain, which is enclosed within the skull, and the spinal cord, which is enclosed within the vertebral column.

THE BRAIN

The brain (see Fig. 16/2) is composed of:

(a) Fore-brain; the cerebrum, composed of the right and left cerebral hemispheres
(b) Mid-brain; the cerebral peduncles and the corpora quadrigemina
(c) mind-brain; the pons, the medulla oblongata and the cerebellum.

The Cerebrum
The cerebrum, the largest part of the brain, is composed of a right and left cerebral hemisphere. The surfaces of the cerebral hemispheres are marked by many irregular folds called gyri, which are separated by fissures called sulci. About four-fifths of the total amount of cortex is hidden in the sulci. The number and position of the gyri is fairly constant.

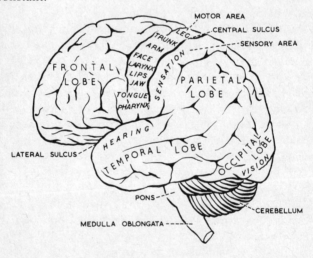

Fig. 16/2 Lateral view of brain, and its functional areas

A deep *longitudinal sulcus* separates the two hemispheres, completely at the front and back, but in the middle only as far down as the *corpus callosum*, a thick and broad band of nerve-fibres connecting the two hemispheres. The *central sulcus* (fissure of Rolando) runs downwards and slightly forwards from about the middle of the top of each hemisphere, between the motor and sensory parts of the cortex. The *lateral sulcus* (fissure of Sylvius) winds round the outer surface of each hemisphere.

Each cerebral hemisphere is divided into lobes named after the bone overlying them:

(a) The *frontal lobe*: the part in front of the central sulcus and above the lateral sulcus.

(b) The *parietal lobe*: behind the central sulcus and above the posterior end of the lateral sulcus.

(c) The *occipital lobe*: at the back of the cerebral hemisphere and behind the parietal lobe.

(d) The *temporal lobe*: below the lateral sulcus.

Functions of the Cerebrum

1. *Motor and premotor areas* The motor and premotor areas are situated in the frontal lobe. The motor area occupies a vertical gyrus immediately in front of the central sulcus, and the premotor area occupies a strip of cortex in front of the motor area. In these areas the body is represented upside down – with a large area for the head and face at the bottom, a large area for the hand above that for the face, and then smaller areas for the arm, trunk, leg and perineum (the area between the thighs). About half the motor cells from which arise the motor fibres are situated in the motor areas, the rest in other parts of the cortex.

2. *Sensation*   At the highest levels the sensations from the skin are appreciated in the anterior part of the parietal lobe, immediately behind the central sulcus, in a vertical gyrus in which the body is also represented upside down, with a large area for the head and face at the bottom.

Gross sensations of pain and temperature are appreciated in the thalamus, below the cortical level, and finer sensations only are transmitted to the cortex.

3. *Sight* Shapes, colour and movement are recognised in the visual area at the tip of the occipital lobe. The visuopsychic area, in which visual impressions are stored and associations are made, is immediately in front of the visual area.

4. *Hearing* Sounds are appreciated and loudness and pitch are determined in the auditory area in the superior temporal gyrus. The psycho-auditory area, in which sounds are stored, interpreted and associated, surrounds the auditory area.

5. *Speech* Speech in its medical sense includes all activities with words – speaking, listening, writing, reading. A speech centre, which directs and controls all these functions, is present in the left cerebral hemisphere for all right-handed people and for most if not for all left-handed people. It occupies part of the cortex of the temporal lobe and of the frontal and parietal lobe on either side of the bottom of the central sulcus. This speech centre has many connections with the parts of the brain involved in movements of the head and hand, and with the auditory and visual areas of the brain.

6. *Smell and taste* These are appreciated in areas deep in the temporal lobe.

The *association areas* of the cortex are those which do not appear to have any limited function, but are thought to be involved in higher cerebral activities – thinking, learning, remembering; and in them the richness of the associations of one part with another by their nerve fibres seems to be as important as the number of cells. The prefrontal lobes, those parts of the frontal lobes in front of the premotor area, have been thought to be the areas in which abstract thought, alertness and initiative are centred.

The *cells of the cortex* number many millions and are arranged in five or six distinct layers. The white matter below the grey cortex is composed of:

1. Fibres connecting different parts of the cortex on the same side.
2. Fibres passing from one cerebral hemisphere to the other through the corpus callosum.
3. Fibres connecting the cortex with the basal ganglia, mid-brain, hind-brain and spinal cord.

The *basal ganglia* are several large rounded masses of grey matter embedded deep inside the cerebral hemispheres (see Fig. 16/3).

1. The *thalamus* is a relay station in the path of the sensory nerve-fibres as they pass upwards towards the cortex from the spinal cord and lower levels of the brain. Gross sensations, of pain and changes in temperature, are appreciated in it.
2. The *corpus striatum* (which includes two large masses of grey matter called the caudate and lenticular nuclei) is probably the source of a primitive motor pathway, with fibres running to lower parts of the brain and to the spinal cord and with its functions normally dominated by the motor cortex of the cerebral hemispheres.

The *hypothalamus* is a small area of grey matter in the floor of the third ventricle. It lies immediately above the pituitary gland, to which it is connected by a short stalk and capillaries. It secretes hormones which pass into the pituitary gland.

The Mid-Brain

The mid-brain is small, about 2cm long, and overgrown by the fore-brain. It consists mainly of two cerebral peduncles, composed of nerve-fibres passing up to or down from the cerebrum, and at the back of four corpora quadrigemina, rounded projections of nerve cells, the upper pair being concerned with sight reflexes, the lower pair with hearing.

The Hind-Brain
The hind-brain consists of the pons, medulla oblongata and cerebellum.

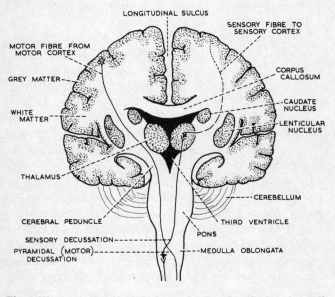

Fig. 16/3 A section through the brain, showing the basal ganglia and the course of the main motor and sensory tracts

The *pons* extends from the mid-brain above to medulla oblongata below. It consists mainly of motor and sensory fibres and of fibres with connections in the cerebellum.

The *medulla oblongata* is a much narrower part of the brain and extends from the pons above to the top of the spinal cord below. It is composed of:

(a) Motor and sensory nerve-fibres. The *pyramidal decussation*, which occurs in the medulla oblongata, is the visible crossing of many of the motor fibres from one side to the other. As a result of this crossing the right cerebral hemisphere controls muscular movements on the left side of the body, and the left cerebral hemisphere movements on the right side.

(b) Nuclei from which arise several of the cranial nerves.

(c) The cardiac centre which controls the heart and the respiratory centre which controls respiration.

The term *brain stem* is used to describe the midbrain, pons and medulla oblongata as a functioning unit. The *reticular system* is a system of nerve-cells and fibres which extends a long way deep in the brain stem. It is concerned in the production of consciousness and unconsciousness. The *cerebellum* controls balance and muscle co-ordination.

THE CRANIAL NERVES

The cranial nerves are twelve pairs of nerves which arise from the brain and are numbered from before backwards. Some have sensory fibres, some motor fibres, and some both sensory and motor fibres. They perform important functions in the head and neck, and the tenth (vagus) nerve also extends into the chest and abdomen.

1. Olfactory Nerves: smell.
2. Optic Nerves: sight.
3. Oculomotor Nerves: movements of the eyeballs, size of the pupils, accommodation in the eyes.
4. Trochlear Nerves: movements of the eyeballs.
5. Trigeminal Nerves: motor fibres to the muscles of mastication; sensory fibres from face, nose, mouth; taste.
6. Abducens Nerves: movements of the eyeballs.

7. Facial Nerves: motor nerves to the muscles of expression in the face.
8. Auditory Nerves: hearing; balance.
9. Glossopharyngeal Nerves: raising the larynx in swallowing; sensation in the pharynx and tongue; taste.
10. Vagus Nerves: movements of swallowing and phonation; parasympathetic functions in the lungs, heart, alimentary tract, etc.
11. Accessory Nerves: movements of the head on the shoulders.
12. Hypoglossal Nerves: movements of the tongue.

THE SPINAL CORD

The spinal cord is about 45cm long and occupies the upper two-thirds of the vertebral canal, from the atlas as far down as the first or second lumbar vertebra (see Fig. 5/4, p. 43). Above it is continuous with the medulla oblongata. It is cylindrical, but in the cervical and lumbar regions it shows swellings because of the additional cells and fibres required for the arms and legs. At its bottom end it tapers suddenly and is attached by a fibrous filament to the coccyx.

The anterior and posterior roots of its spinal nerves emerge from its sides (see Fig. 16/4). Because of the shortness of the spinal cord, the lumbar and sacral nerves have to pass for some distance downwards within the vertebral canal, being collectively called the cauda equina (horse's tail) before they can enter their appropriate holes in the side of the vertebral column.

The grey matter of the cord is on the inside, in the shape of an H, the projections in front being called the anterior horns and those behind the posterior horns. As elsewhere in the nervous system, this grey matter is composed of nerve-cells. The rest of the spinal cord is white matter composed of myelinated motor and sensory nerve fibres.

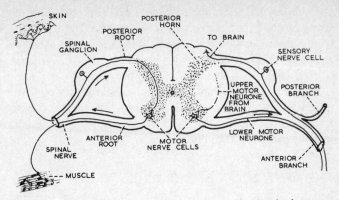

Fig. 16/4 A section through the spinal cord and spinal nerves, showing on the left the structures involved in a simple reflex action

The Spinal Nerves

The spinal nerves supply the parts of the body not supplied by the cranial nerves. There are 31 pairs:

cervical nerves	8
thoracic nerves	12
lumbar nerves	5
sacral nerves	5
coccygeal nerve	1

Each is attached to the spinal cord by an anterior (motor) root and a posterior (sensory) root. Each posterior root has on it a ganglion, the spinal ganglion, in which are the cells of the sensory root fibres. Just beyond this ganglion the anterior and posterior roots fuse to form a spinal nerve, which is therefore a mixed nerve, one which contains both motor and sensory fibres.

Each spinal nerve passes out of the vertebral canal by going through its intervertebral foramen and emerges at the side of the vertebral column. Each then divides into two branches, both of which contain motor and sensory fibres. The posterior branch goes to the skin and muscles at the back of the body, the much larger anterior branch goes to the sides and front of the body.

The cervical, brachial, lumbar and sacral plexuses are groupings of nerves from the anterior branches of the spinal nerves, and from them nerves are distributed to various parts of the body and to the limbs.

THE VENTRICLES OF THE BRAIN

The ventricles of the brain are cavities formed by enlargements and modifications of the original canal that ran down the neural tube when the nervous system was in an early stage of development. They are filled with cerebro-spinal fluid.

The two *lateral ventricles*, right and left, lie deep within the cerebral hemispheres and are irregular in shape, with projections into the frontal, occipital and temporal lobes (see Fig. 16/5). They communicate with one another and with the third ventricle. The *third ventricle* lies centrally and lower than the lateral ventricles and between the thalami; it communicates at its posterior end with the fourth ventricle through the aqueduct, a narrow tube. The *fourth ventricle* is lozenge-shaped and lies behind the pons and medulla oblongata and in front of the cerebellum. In its roof are three tiny openings through which the cerebro-spinal fluid passes out of the ventricular system into the subarachnoid space.

The Cerebro-Spinal Fluid

The cerebro-spinal fluid (CSF) occupies the ventricles and the subarachnoid space. It is formed by the choroid plexuses, which are clumps of capillaries lying within the ventricles, the largest being in the lateral ventricles where most of the cerebro-spinal fluid is formed. It is a clear fluid, resembling in its chemical content the plasma of the blood but differing from plasma in containing little protein. About 500ml are secreted daily, and the amount present at any one time is 120ml. It passes through the ventricular system and leaves the fourth ventricle through the three holes in its roof. It then enters the subarachnoid space and circulates round the outside of the brain and spinal cord. Finally it is absorbed back into the blood in the venous sinuses by passing through some small structures called the arachnoid villi.

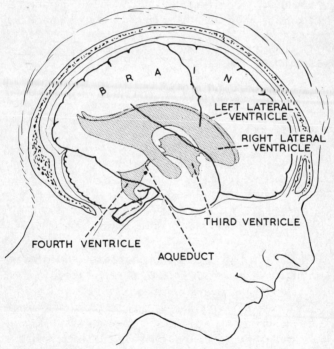

Fig. 16/5 The position of the ventricles within the brain

The functions of the cerebro-spinal fluid are:

1. To maintain a constant volume within the skull by increasing or decreasing in amount with a decrease or increase of the other cranial contents (e.g. an enlargement of the blood vessels).
2. To act as buffer protecting the brain from a blow or jerk.
3. To receive waste products of metabolism from brain tissue and transfer them to the blood.

THE MENINGES

The meninges are three membranes which enclose the brain and spinal cord (see Fig. 16/6). From within outwards they are the pia mater, the arachnoid membrane and the dura mater.

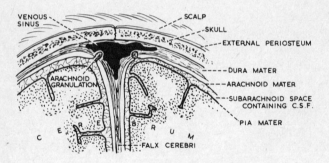

Fig. 16/6 The skull and meninges and a venous sinus

The *pia mater* is a thin, vascular membrane, which covers the entire surface of the brain and spinal cord, dipping down into the fissures between the gyri.

The *arachnoid membrane* is a delicate membrane which is in some places fused with the pia mater. In some places it is separated from it by the subarachnoid space, which is filled with cerebro-spinal fluid. The subarachnoid membrane and space extend a little way below the lower end of the spinal cord, and this makes possible the operation of lumbar puncture, in which a needle is inserted between two of the lumbar vertebrae into the subarachnoid space and a specimen of cerebro-spinal fluid is withdrawn without injuring the spinal cord.

The *dura mater* is a thick, stiff membrane of fibrous tissue lining the interior of the skull (where it forms the internal periosteum) and protecting the brain from injury. It also forms the falx cerebri, a strong vertical plate between the two cerebral hemispheres, and the tentorium cerebelli, a horizontal sheet separating the occipital lobes and the cerebellum.

The *venous sinuses* of the skull, into which passes blood from the brain and skull, are formed of layers of dura mater.

The Blood Supply of the Brain

The brain is supplied with blood through the two internal carotid arteries and the two vertebral arteries.

The *internal carotid artery* on each side arises in the neck from the common carotid artery and passes into the cranial cavity through a hole in the temporal bone. The *vertebral artery* on each side comes from the subclavian artery low in the neck, passes upwards through the holes in the lateral processes of the cervical vertebrae, and enters the cranial cavity through the foramen magnum in the occipital bone. Within the skull the two vertebral arteries unite to form one basilar artery. The circle of Willis at the base of the brain is a circle of arteries connecting the two internal carotid arteries in front with the basilar artery behind (see Fig. 16/7).

The cerebral hemispheres are supplied by anterior, middle and posterior cerebral arteries. From them arise two sets of vessels:

(a) superficial arteries, which run in the pia mater on the

surface of the cerebral hemispheres and supply the cortex and parts lying immediately below the cortex.

(b) perforating arteries which pass deep into the brain to supply the basal ganglia and other deep-lying structures.

The pons, medulla oblongata and cerebellum are supplied by branches from the vertebral and basilar arteries.

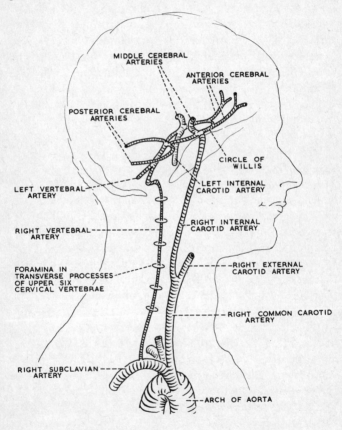

Fig. 16/7 The blood supply to the brain

The venous blood from the brain passes into large venous sinuses close to the skull. From them most of the blood passes into the right and left jugular veins.

The Electro-Encephalogram

The electro-encephalogram (EEG) is a recording of the electrical discharges produced by the simultaneous activity of millions of brain-cells. The discharges are recorded through electrodes placed on the scalp and are magnified to a size large enough to be recorded and studied (see Fig. 16/8). Changes in electrical potential appear in the form of waves. The *alpha rhythm* is the normal pattern of slightly irregular small waves occuring at the rate of 8–13 a second and produced in both cerebral hemispheres. The waves begin in childhood and remain constant for an individual. They are best seen when a person is relaxed and has his eyes shut. Larger abnormal waves appear in some abnormal conditions of the brain.

Sensory and Motor Tracts

Sensory tracts   The sensations we feel in the skin are transmitted as sensory impulses through the spinal and some of the cranial nerves to the central nervous system. Sensory nerves form the posterior roots of the spinal cord, their cells being in the spinal ganglion on each root. Once within the spinal cord the fibres can (a) form a synapse with the cells of the motor fibres in the anterior horn of grey matter, through which simple spinal reflexes can take place; (b) communicate with segments of the cord above and below their own; or (c) pass up the cord to nuclei in the medulla oblongata, whence they are relayed to the thalamus and from the thalamus to the sensory cortex.

All these sensory fibres at some point in their course pass to the other side, so that sensations affecting one side of the body are appreciated in the other side of the brain.

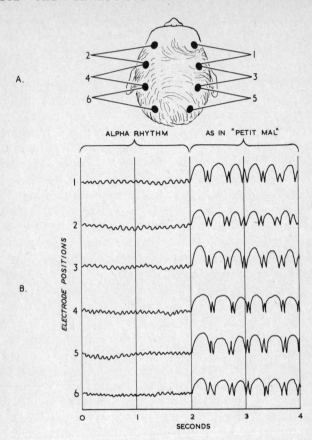

Fig. 16/8 An electro-encephalogram, showing normal rhythm and an abnormal rhythm of spikes and waves, typical of petit mal epilepsy

Motor tracts   Motor impulses which stimulate muscles to contract or relax and are conveyed through two sets of neurones – the upper motor neurones which convey impulses from the brain to the spinal cord and the lower motor neurones which convey impulses from the spinal cord to the muscles.

The *upper motor neurones* begin in cells in the motor area of the cerebral cortex or in other parts of the cortex. The fibres pass through the cerebral hemispheres, the cerebral peduncles of the mid-brain, and then through the pons to the medulla oblongata. In the medulla many of them cross to the other side in the pyramidal decussation (those which do not cross here cross lower down), and then run down the spinal cord to end by forming synapses with cells in the anterior horn of grey matter, where the impulses they convey are handed over to the lower motor neurones.

The *lower motor neurones* begin in cells situated in the anterior horn of grey matter in the spinal cord. Their fibres emerge as the anterior roots of the spinal cord, pass into the spinal nerves and end in the voluntary muscles. Each fibre divides into smaller branches, and the smallest branches end in tiny plates which lie on the muscle fibres. There is a chemical transmission of impulse from the nerve plate to the muscle fibre.

One of the important differences in function between the upper and the lower motor neurones is that the upper are concerned with movements as a whole and the lower more with the stimulation of the individual muscles needed to make these movements.

Reflex Actions
Reflexes can be simple (unconditioned) or conditioned.

A *simple (unconditioned) reflex* is the automatic response of a motor organ to a sensory stimulation. The majority of the activities of the nervous system are carried out by reflex actions of which we are unaware. Typical of reflex actions are the changes that take place in the pupil of the eye when it is exposed to different intensities of light. If someone walks out of a dark room into sunlight, his pupils promptly constrict so that less light enters the eye; and for this to be done, the stimulus of bright light falling on the retina has to be transmitted to the brain stem and motor stimuli have to be transmitted to the muscles in the eye which control the size of the pupil. If a person goes from a bright light into a dim one, a different set of reflex responses dilate the pupil.

A *spinal reflex* involves a sensory nerve, the spinal cord and a motor nerve. A painful stimulus will produce a muscular contraction which jerks the part away from the stimulus. In testing the knee jerk which is a spinal reflex, the doctor taps the patella tendon below the knee; this lengthens it slightly, the quadriceps muscle in the thigh is stimulated, and the contraction of it causes the jerk.

Simple reflexes are inborn and more or less the same in all people. Similar reflexes occur with glands, e.g. the smell and taste of food reflexly stimulate the secretion of saliva. The entry of food into the stomach reflexly produces a secretion of gastric juice.

A *conditioned reflex* is a reflex in which the response is produced by some stimulus other than the natural one. It is built upon a natural one. Ringing a bell would not normally produce salivation, but if the ringing always immediately precedes the giving of food to an animal, it will in time by itself and without the appearance of food produce salivation.

THE AUTONOMIC NERVOUS SYSTEM

The autonomic nervous system is the part of the nervous system concerned with the automatic control of internal organs, e.g. the rate of the heart, the size of arterioles, the activities of the alimentary tract, the secretion of adrenaline. It functions independently of the rest of the nervous system, but it is connected with the brain and

THE NERVOUS SYSTEM 165

spinal cord by certain nerves and it is controlled by centres in the hypothalamus and medulla oblongata. It consists of two parts: the sympathetic nervous system and the para-sympathetic nervous system, whose actions are usually antagonistic – where one stimulates, the other relaxes.

The Sympathetic Nervous System

The sympathetic nervous system consists of (a) two chains of nerve fibres and ganglia, which run down the side of the vertebral column from neck to sacrum, (b) nerves which leave these ganglia, (c) ganglia close to the organs they serve, and (d) nerve fibres passing from these ganglia to the organs.

Nerve fibres of the sympathetic system pass to the heart, the muscular walls of arteries and arterioles, the lungs, the alimentary tract, the kidneys, the bladder, the sexual organs, the salivary glands and other tissues. Near the organs they serve are four great plexuses of nerves and ganglia: the cardiac plexus in the chest, the coeliac (solar) plexus in the upper abdomen, the hypogastric plexus in the lower abdomen, and the pelvic plexus in the pelvis.

The Parasympathetic Nervous System

The parasympathetic nervous system is in two parts, a cranial and a sacral part, and uses cranial and sacral nerves to distribute its fibres.

The *cranial section* supplies fibres to the pupil of the eye, the lacrimal and salivary glands, and via the tenth cranial (vagus) nerve the larynx, heart, lungs, alimentary tract as far as the transverse colon, liver, spleen and kidneys.

The *sacral section* supplies the rest of the large intestine, the bladder and genitalia.

The Functions of the Autonomic Nervous System

Some important functions of the two parts of the auto-nomic nervous system are summarised in this table:

Organ	Sympathetic system	Parasympathetic system
Heart	Rate and output increased	Rate and output decreased
Arteries to muscles	Dilated	Constricted
Arteries to abdominal organs	Constricted	
Blood pressure	Raised	Lowered
Peristalsis in alimentary tract	Decreased	Increased
Sphincters of alimentary tract	Closed	Relaxed
Liver	Glycogen converted into glucose	Glucose converted into glycogen
Adrenal glands	Increased secretion of adrenaline	
Skin	Hairs erected, and sweat glands stimulated	

From this table it will be appreciated that the reactions of the sympathetic system are directed towards the mobilisation of the resources of the body to meet a danger or emotional crisis. The heart beats faster and pumps out more blood, the muscles get more blood with which to carry out their functions and the other organs get less, the movements of the alimentary tract are diminished and the closing of their sphincters stops excretion. The parasympathetic system has the opposite effects: its reactions are those of quiet life and the conservation of energy; under its stimulation activities are reduced, the heart beats more slowly, the blood pressure drops, the processes of digestion are pro-moted.

In normal life the activities of the internal organs are kept balanced by the interaction of the two systems.

Chapter 17

THE SPECIAL SENSES

VISION

The Eye

The eye lies in a cushion of fat within the protection of the bony walls of the orbit. The front of it is protected by the eyelids, the eyebrows and a circular muscle which surrounds the orbit in front. At any threat to the eye the muscle fibres are reflexly contracted, the eyebrows drawn down, and the eyelids closed.

The eyeball is almost spherical. In front the cornea, the eye's transparent window, bulges slightly forwards like the glass of a watch. The eyeball contains several structures and has three coats: (a) the cornea and sclera, (b) the iris, ciliary muscle and choroid coat, and (c) the retina (see Fig. 17/1).

The *cornea* and *sclera* form an outer fibrous coat. The *cornea* is the transparent front of the eye through which light enters. It has no blood vessels and gets its nutrition from the aqueous humour just behind it. The *sclera*, the 'white' of the eye, forms an outer opaque coat around the rest of the eyeball. The muscles that move the eyeball are inserted into it just behind the ring where cornea and sclera fuse. The optic nerve pierces it at the back of the eyeball.

The *iris*, the *ciliary body* and the *choroid coat* form the vascular and muscular coat of the eyeball. The *iris* is a coloured circular diaphragm surrounding the pupil, the circular hole at its centre, and lies a little distance behind the cornea and just in front of the lens. It contains two sets of involuntary muscle fibres: a circular set which contracts the pupil and a radiating set which dilates it. The colour of the iris varies with the amount of pigment in it; the more pigment the darker the iris. The *ciliary body* is a ring of vascular and muscular tissue behind the iris and continuous with its outer margin. It contains the ciliary muscle, which is attached to the suspensory ligament of the lens. This ligament encloses the lens, and when the ciliary muscle contracts, the ligament relaxes and this relaxation allows the lens to become more convex. The *choroid coat* is a vascular membrane between sclera and retina.

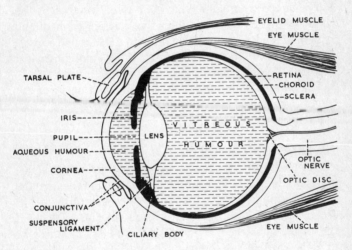

Fig. 17/1 The structure of the eye – lateral view

The *retina* is the nervous coat of the eye and is a thin delicate membrane composed of cells capable of reacting to light. In front it ends just behind the ciliary body. The optic nerve pierces it at the back at a spot called the optic disc; this optic disc, being composed solely of nerve-fibres, is insensitive to light and hence called the blind spot. The small artery and vein which supply the retina reach it by passing through the centre of the disc.

The *rods* and *cones* are the light-sensitive cells of the retina. The rods are long and thin, the cones short, fat and pointed. They are neurones, and through a second set of cells communicate with a third set, from which arise the fibres of the optic nerve.

The contents of the eyeball are:

1. The *lens* which is a biconvex, elastic and transparent disc situated immediately behind the iris and pupil and held in position by the suspensory ligament.
2. The *aqueous humour* which is a fluid secreted by the ciliary processes. It occupies the space in front of and behind the iris, supplies nutrition to the lens and cornea, and maintains the pressure within the eyeball constant at about 25mm of mercury.
3. The *vitreous humour* which is a jelly that occupies the posterior four-fifths of the eye, behind the lens and ciliary body.

The External Muscles of the Eyeball

The eyeball is moved by six muscles: four recti (straight) muscles and two oblique muscles. The recti muscles and the superior oblique muscle arise from the back of the orbit; the inferior oblique arises from the inferior surface of the orbit, near the front. All the muscles are inserted into the eyeball just behind the junction of cornea and sclera.

Other Structures

The *eyelids* contain a stiff plate of connective tissue called the tarsal plate and fibres of the muscle encircling the eye. The upper eyelid can be raised by a muscle coming from the back of the orbit and over the top of the eyeball and upper rectus muscle. The *conjunctive* is the thin membrane which lines the inner surface of the eyelids and is continued over the front of the cornea.

The *lacrimal gland*, which secretes the tears, is situated in the orbit just above and to the outer side of the eyeball (see Fig. 17/2). The tears it secretes pass in a thin film over the front of the conjunctiva. Some evaporate on the way and the rest passes through tiny, just visible holes at the inner end of each eyelid and then down a naso-lacrimal duct, which at its lower end opens into the nose – which is why we have to blow our nose when we cry.

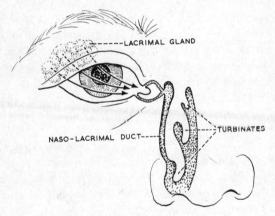

Fig. 17/2 The lacrimal system

The Optics of the Eye

In basic construction an eye is like a camera. Light enters it through a hole, the pupil, in front. The amount of light entering it is controlled by a diaphragm, the iris, and is focused by a lens on to the back of the eye, where it stimulates a sensitive medium, the retina. We can regard the eye as a wide-angled cine-camera, capable of taking an unlimited number of colour photographs throughout the life of its owner.

Light enters the eye by passing through the cornea, aqueous humour, lens and vitreous humour. All these four cause it to be refracted, i.e. bent out of its original direction. The lens is the most important of these refracting media. The degree of its convexity is altered by contraction or relaxation of the ciliary muscle, acting through the suspensory ligament. The lens becomes more convex when near objects are looked at and less convex when distant objects are looked at. Accommodation is the word used to describe this alteration in refractive power, by which visual impressions are maintained in focus on the retina.

The amount of light entering the eye is controlled by the action of the iris. When the light is bright, the pupil is constricted; when the light is dim, the pupil is dilated.

In addition to causing refraction, the lens inverts the picture so that an inverted image falls on the retina. There the rods and cones are stimulated. The rods are concerned with seeing in poor light, the cones with accuracy of vision and with colour vision.

The rods and cones have the ability to turn light into nervous impulses. These impulses are transmitted in the fibres of the optic nerve to the midbrain, whence they are relayed to the visual area in the occipital cortex on both sides of the brain (see Fig. 17/3).

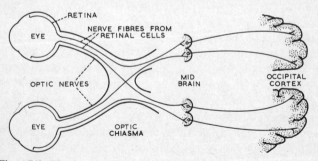

Fig. 17/3 The optic nerves, the optic chiasma, and optic tracts

The cones contain a pigment called visual purple, which on exposure to light becomes 'bleached' and has to be regenerated. Dark adaptation (the ability to see better in the dark after a time) is due to regeneration of visual purple, which occurs in darkness or when the eyes are shut. Vitamin A is necessary for this regeneration, and an inability to adapt to the dark is an early symptom of lack of vitamin A.

The advantages of having two eyes and therefore binocular vision are: (a) defects in the visual impression in one eye (e.g. the blind spot) are not noticed; (b) the slight obstruction to vision that the nose causes is not noticed; (c) stereoscopic vision is achieved, and (d) distances and sizes can be judged more accurately.

HEARING

The ear consists of three parts – the outer ear, the middle ear, and the inner ear (see Fig. 17/4).

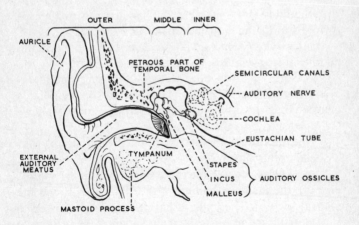

Fig. 17/4 The structure of the ear

The Outer Ear

The outer ear consists of the auricle and the external ear. The *auricle* is a plate of fibro-cartilage covered with skin and connected to the skull by ligaments and small muscles. The *external auditory meatus* is the tube from the auricle to the tympanic membrane. It is about 2.5cm long and forms a S-shaped curve, which can be straightened by pulling the auricle backwards and upwards. The outer third is composed of cartilage, the inner two-thirds of bone. The whole tube is lined with skin, in which are hairs and the glands that produce wax.

The Middle Ear

The middle ear is a small space in the temporal bone. It is lined with mucous membrane and contains three small bones called the auditory ossicles.

The *tympanic membrane* (ear drum) separates the external auditory meatus from the middle ear. It is a taut, oval fibrous membrane, set obliquely at an angle of about 55° with the floor of the meatus. The handle of the malleus is attached to its inner surface as far as its centre.

The *pharyngo-tympanic (Eustachian) tube* connects the middle ear to the naso-pharynx. Its function is to keep the air pressure equal on both sides of the tympanic membrane; and its bottom end, where it enters the pharynx, is guarded by a valve, which is opened by the act of swallowing to let air up to the middle ear.

At the back of the middle ear a hole communicates with air-cells in the mastoid process of the temporal bone.

The *auditory ossicles*, the three small bones in the middle ear, are the malleus (hammer), the incus (anvil) and the stapes (stirrup). The malleus has one process attached to the tympanic membrane; the incus connects the malleus to the stapes; the stapes has a flat plate fixed in a small hole in the wall of the inner ear; and in this way the ossicles bridge the gap between the tympanic membrane and the inner ear. Through them the vibrations of the membrane are transmitted across the middle ear to the inner ear.

The Inner Ear
The inner ear is a complicated structure in which are combined the organ of hearing and the organ of balance.

(a) *Hearing*   The *cochlea* is the essential part of the organ of hearing. It is a bony tube within the temporal bone and wound spirally around a spike of bone, through which pass branches of the auditory nerve. It is lined with membrane and filled with a fluid called perilymph. Two other membranes run the length of the spiral, enclosing another fluid called endolymph, and to one of them, the basilar membrane, is attached the *organ of Corti* (spiral organ), a structure of special cells, in which begin the fibres of the auditory nerve. These fibres run to nuclei in the brain, where synapses are formed with fibres running to the auditory centre in the superior gyrus of the temporal lobe.

(b) *Balance*   The three *semicircular canals*, immediately behind the cochlea and in close anatomical contact, are concerned with balance and the position of the head, and have nothing to do with hearing.

The Mechanism of Hearing
Sound is a form of wave-motion, varying in amplitude (which decides the loudness of the sound) and frequency (which decides the pitch). Sounds of high frequency have a high pitch and sounds of low frequency a low pitch, and the human ear can hear sounds of frequencies of 30–15 000 vibrations a minute.

Sound waves strike the tympanic membrane, setting up vibrations in it, the conical shape of the membrane enabling it to respond to a wide range of pitch. The movements of the membrane are transmitted through the ossicles in the middle ear and so to the fluids in the inner ear. The basilar membrane is tight and short in its lower part and gradually becomes longer and less tight towards its upper part. The short, tight parts are set vibrating by notes of high pitch, and the longer, less tight parts by notes of low pitch. The movements of the basilar membrane are communicated to the nerve-endings in the organ of Corti, and through them nervous impulses are transmitted to the auditory centres in the temporal lobes of the brain, where they are appreciated as sounds.

SMELL AND TASTE

The senses of smell and taste are feebly developed in man, the cortical areas for them being small.

The *organ of smell* is the olfactory area in the roof of the nasal cavities, from which fibres of the olfactory (first cranial) nerves pass through holes in the ethmoid bone of the skull into the cranial cavity. There they enter the olfactory bulbs, which are swellings at the ends of the olfactory tracts. The tracts convey the impulses to the brain.

The *organ of taste* is composed of tiny taste-buds in the surfaces of the tongue and cheeks. From taste-cells in these buds nerve fibres run to the brain in the fifth (trigeminal) and ninth (glossopharyngeal) cranial nerves. Probably only four tastes can be detected by taste-buds – sweet, salt, sour and bitter, in that order from the tip of the tongue backwards. Any other appreciation of taste is a mixture of smell, sight, feeling and heat; and a person blindfolded and with nose pinched to stop him from smelling, cannot easily tell what food is put into his mouth.

Chapter 18

THE ENDOCRINE GLANDS

The endocrine glands are glands that pass their secretions directly into the blood and not down a duct. These secretions are called *hormones*, and they are essential for many of the activities of the body. They are carried in the blood to all parts of the body, and commonly one hormone has an effect on the secretions of others. Some organs (e.g. the pancreas) have other functions as well as the production of hormones.

The following are known to produce hormones: the hypothalamus-pituitary complex, the thyroid gland, the parathyroid glands, the adrenal glands, the pineal gland, the alimentary tract, the pancreas, the testes, the ovaries, the placenta and the kidneys.

THE HYPOTHALAMUS-PITUITARY COMPLEX

The *hypothalamus* is a small area of cells in the base of the brain. It is connected by a stalk of nerve fibres and capillaries with the pituitary gland, which lies immediately below it.

The *pituitary gland* is a rounded body, about the size of a small cherry, lying in the pituitary fossa of the sphenoid bone of the skull and connected by a stalk with the hypothalamus (see Fig. 18/1). It is composed of two lobes, anterior and posterior, differing in origin and structure.

The hypothalamus produces a number of hormones, which pass down the stalk to the pituitary gland, where they

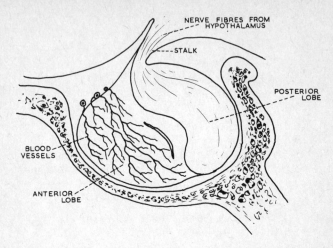

Fig. 18/1 The pituitary gland and pituitary fossa

are stored until required by the body (see Fig. 18/2). The pituitary gland produces one hormone of its own.

1. *Growth hormone* (GH) causes nitrogen to be retained in the body and is essential for growth. Over-production of it in childhood and adolescence causes the person to become very tall, under-production causes him be short or dwarfed.

2. *Thyro-trophic stimulating hormone* (TSH) stimulates the thyroid gland to produce its hormones, thyroxine and tri-iodothyronine. It is produced by the hypothalamus and liberated from the pituitary gland by a fall in the amount of thyroxine in the blood.

3. *Adreno-cortico-trophic hormone* (ACTH) stimulates the cortex of the adrenal gland to produce glucocorticoids.

4. *Sex gland hormones (gonadotrophic hormones)*: (a) *Female . Follicle-stimulating hormone* (FSH) causes ripening of the ovarian follicles. *Luteinising hormone (*LH*)* combines with FSH to complete the ripening of ovarian

follicles and stimulates the development of the corpus luteum.

(b) *Male. Interstitial-cell stimulating hormone* (ICSH) stimulates the interstitial cells of the testes to produce androgens. It is identical with the female LH.

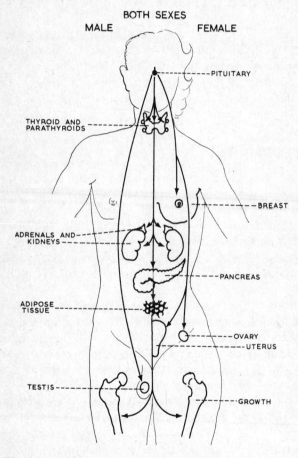

Fig. 18/2 A diagram of the functions of the pituitary gland

5. *Melanocyte stimulating hormone* causes pigmentation of the skin by stimulating the production of the pigment melanin.
6. *Antidiuretic hormone* stimulates the cells of the tubules of the kidney to reabsorb water from the fluid flowing through them and is so involved in the production of urine.
7. *Oxytocin* stimulates contraction of the uterus after childbirth and contractions of the muscle in the ducts of the breast, driving the milk along them.
8. *Prolactin* (the only hormone actually produced in the pituitary gland) is involved in the production of milk in the female breast.

THE THYROID GLAND

The thyroid gland is an H-shaped gland in the front of the neck and consists of two conical lobes connected by a narrow portion called the isthmus. The lobes lie on either side of the larynx and trachea, the isthmus lies across the front of the top of the trachea (see Fig.18/3). The gland is composed of many completely enclosed tiny vesicles, small globules containing a semi-fluid, hormone-containing colloid.

Thyroxine and *tri-iodothyronine* are hormones which are essential for normal metabolism. They contain iodine, which the cells of the gland can take out of the blood. Their production is stimulated by the TSH of the hypothalamus-pituitary complex, the TSH being produced in response to a fall of the amount of thyroxine in the blood.

Thyro-calcitonin is a hormone which reduces the amount of calcium in the blood, its action being the opposite of that of parathyroid hormone produced by the parathyroid glands. It is produced when the amount of calcium in the blood rises too high.

THE PARATHYROID GLANDS

The parathyroid glands are four small glands which lie behind the thyroid gland or are embedded in the back of it (see Fig. 18/3). There are usually an upper and a lower pair of them, but they can vary in number and position. They are composed of clumps of cells surrounded by sinusoids.

Parathyroid hormone raises the amount of calcium in the blood by: (a) promoting the absorption of calcium by the intestine, (b) transferring calcium from bone into blood, and (c) increasing the absorption of calcium by the tubules of the kidney from the fluid in them. When the blood calcium starts to become too high, the secretion of parathyroid hormone is reduced and that of thyro-calcitonin by the thyroid gland is increased.

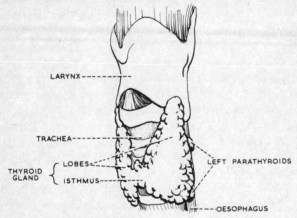

Fig. 18/3 The thyroid and parathyroid glands

THE ADRENAL GLANDS

The adrenal glands are two glands, one on each side, at the back of the abdomen, capping the upper pole of each kidney and embedded in the fat which surrounds the kidneys (see Fig. 10/4, p. 106). Each consists of two parts, a cortex and a medulla, anatomically in close contact but different in origin, structure and function.

Adrenal Cortex

The cortex is on the outside of the gland and is yellow. It consists of several layers of cells and contains much cholesterol and vitamin C. It produces:

1. *Glucocorticoids* – hydrocortisone and others. Their secretion is regulated by ACTH from the hypothalamus-pituitary complex. Hydrocortisone (a) is an antagonist of insulin, causes glycogen to be deposited in the cells of the liver, and raises the blood sugar; (b) breaks down tissue proteins, which are converted in the liver into glycogen, and (c) is involved in the exchange of water, sodium and potassium between cells and the tissue spaces.
2. *Mineralo-corticoids* Aldosterone, the principal one, regulates the amount of sodium in the body.
3. *Androgens* are produced in males only. They are responsible for the development of secondary male sexual characteristics – growth of hair on the face, deepening of the voice, etc. Their actions are weaker than those of testosterone produced by the testes.

Adrenal Medulla

The adrenal medulla is developmentally and functionally part of the sympathetic nervous system. It is grey in colour and enclosed by the cortex, which has grown round it. It produces:

1. *Adrenaline*, which increases the output and rate of the heart, dilates blood vessels in skeletal muscles, dilates the coronary arteries, constricts blood vessels in the skin and internal organs, decreases peristalsis of the intestine and

contracts its sphincters, dilates the bronchi, increases the consumption of oxygen, and converts glycogen in the liver into glucose.

2. *Nor-adrenaline*, which causes a rise in blood pressure. It has the same action on the intestine as adrenaline. It has little effect on metabolism.

Adrenaline and nor-adrenaline are slightly different chemically and have similar, but not identical actions. They are both secreted in response to stress and enable the body to take effective action in a dangerous situation.

THE PANCREAS

In addition to secreting pancreatic juice, the pancreas produces two hormones – insulin and glucagon. They are produced in the islets of Langerhans, where there are two slightly different kinds of cells – beta cells which produce insulin and alpha cells which produce glucagon.

1. *Insulin* is produced in response to a rise in the amount of sugar in the blood, as after a meal. Its function is to promote the utilisation of glucose by cells and to prevent the blood sugar from rising too high. It causes glucose to be stored as glycogen in the liver. As it is a protein it cannot be used by mouth for the alimentary tract would break it down, and in the treatment of diabetes mellitus it has to be given by injection.

2. *Glucagon* raises the blood sugar by converting glycogen in the liver into glucose. Its action is the opposite to that of insulin, and its production is stimulated by a fall in the blood sugar.

The blood sugar is kept within normal limits by the action on the one hand of insulin and on the other of glucagon and adrenaline.

THE TESTES

In addition to producing spermatozoa the testes produce androgens, the male sex hormones, of which the most important is testosterone. They are produced by the interstitial cells which lie between the spermatozoa-producing tubules.

Testosterone is produced by the interstitial cells by the influence of ICSH from the hypothalamus-pituitary complex. It causes the development of the secondary sexual characteristics (being more powerful in this than the androgens from the adrenal cortex) and builds up proteins from amino-acids.

THE OVARIES

In addition to producing ova the ovaries produce hormones in the ovarian follicles and the corpora lutea.
1. *Oestrogens* are produced in the ripening ovarian follicle and corpus luteum. They are responsible for the development of the female secondary sexual characteristics and for producing changes in the endometrium of the uterus during the menstrual cycle.
2. *Progesterone* is produced in the corpus luteum. It is produced in response to stimulation by LH from the hypothalamus-pituitary complex. It produces changes in the endometrium of the uterus and the development of the breast in pregnancy.

THE PLACENTA

The placenta produces *chorionic gonadotrophin*, a hormone which develops and maintains the corpus luteum during pregnancy. Pregnancy tests are based on the detection of this hormone in the urine.

THE KIDNEYS

In addition to their excretory and other functions the kidneys produce *erythropoietin*, a hormone which stimulates the production of red blood cells. They are stimulated to do this by a fall in the amount of oxygen in the blood and in consequence a demand for more haemoglobin.

THE PINEAL GLAND

The pineal gland is a small pink gland lying in a groove at the back of the mid-brain. In early life it produces *melatonin*, a hormone whose function appears to be to delay the onset of puberty. In adult life it becomes calcified and visible on an X-ray of the head.

THE ALIMENTARY TRACT

The stomach produces *gastrin*, a hormone which stimulates acid secretion by the organ. The duodenum produces *secretin* which stimulates the secretion of pancreatic juice, and *cholecystokinin*, which causes the gall-bladder to contract.

Chapter 19

REPRODUCTION AND HEREDITY

THE MALE GENITAL SYSTEM

The *testes*, the male reproductive glands (see Fig. 19/1), hang between the thighs in the *scrotum*, a pouch of dark, thick, ridged skin.

Each testis is composed of lobes. Each lobe is composed of long, thin, twisted seminiferous tubules, from the cells of whose walls the spermatozoa are formed, and of clumps of interstitial cells between the tubules, cells which form testosterone, the male sex hormone. Each testis is enclosed in three coats: a vascular coat, a fibrous coat and the tunica vaginalis which, covering the testis at the front and sides, forms a totally enclosed bag, into which the testis projects.

In fetal life the testes are formed just below the kidneys; from this position they gradually descend and at about the time of birth arrive in the scrotum. The purpose of this change of position appears to be to remove the glands from the core heat of the abdominal cavity to the cooler scrotum, for spermatozoa cannot develop normally in the greater heat. Occasionally they do not get as far as the scrotum and are then called undescended. The normally descended testes remain small until puberty, when they grow rapidly to adult size.

The *epididymis* is a thin tube, very much coiled upon itself, at the back of the testis. The seminiferous tubules open into its upper end, and the spermatozoa are stored in it. From the epididymis the spermatozoa pass into the *vas deferens*, a much thicker tube, which passes upwards out of the scrotum and outwards along the groin. The vas forms part of the *spermatic cord*, a bundle which also contains the blood-vessels, lymph-vessels and nerves of the testis. About the middle of the groin the vas turns inwards and enters the abdomen through an opening called the inguinal ring – a weak spot in the abdominal wall at which a hernia (rupture) can occur. The vas then runs through the pelvic cavity to the back of the bladder where it ends at the union of the seminal vesicle and the ejaculatory duct.

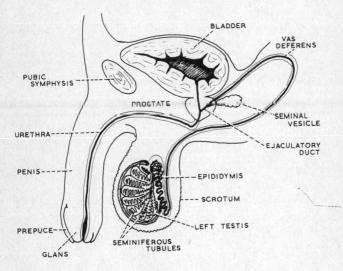

Fig. 19/1 The male genital organs

The *seminal vesicle*, right and left, is a narrow bag on the outer side of the vas and behind the bladder. The vesicles secrete the seminal fluid in which the spermatozoa are ejaculated during sexual intercourse.

The *ejaculatory duct* on each side is a short tube running

from the union of vas deferens and seminal vesicle through the prostate gland. The two ducts open side by side into the urethra, and from this point onwards the male urinary and genital tract are combined.

The *prostate gland* lies just below the bladder, behind the symphysis pubis and in front of the rectum. It is traversed by the urethra and secretes a fluid which is mixed with the fluid produced by the seminal vesicles.

The *penis*, the male sexual organ, is composed of spongy tissue, its shape and size varying with the amount of blood in it. The *urethra* is the tube through which the urine passes from the bladder to the tip of the penis and opens on the glans, the enlarged end of the penis, over which the skin is reflected as the prepuce (foreskin).

THE FEMALE GENITAL ORGANS

The *ovaries*, the female reproductive glands (see Fig. 19/2), are situated in the pelvis, one on each side of the uterus. Each is oval, about 3cm long, and enclosed in peritoneum a little below the uterine tube, to which it is attached. In childhood its surface is smooth, but with repeated ovulation the surface becomes puckered and irregular.

The ovaries have two functions: the formation of ova and the formation of female sex hormones. In both functions they are controlled by hormones from the hypothalamus-pituitary complex.

1. *The formation of ova* Before birth the *germinal epithelium*, a layer of cubical cells, covers the surface of the ovary. Cells from it grow down into the substance of the ovary, some of them to become enlarged into ova, the others to surround each ovum with a layer of cells called the membrana granulosa. At birth each ovary contains several thousand ova, many more than will be needed. No further change takes place until puberty,

when ovarian follicles start to develop. An *ovarian (Graafian) follicle* consists of an ovum and its surrounding cells. From puberty onwards for about thirty years, one mature ovum (occasionally more) is produced every month. Unwanted follicles die. The mature ovum is produced in this way: the ovum enlarges, the surrounding cells multiply and fluid appears between two layers of them; as the follicle develops in this way it moves nearer to the surface of the ovary, touches the surface and ruptures, discharging the ovum from the ovary, the event being called *ovulation*. The open end of the uterine tube has in the meantime been attracted towards the ovary, and the ovum, as it leaves the ovary, passes directly into the tube.

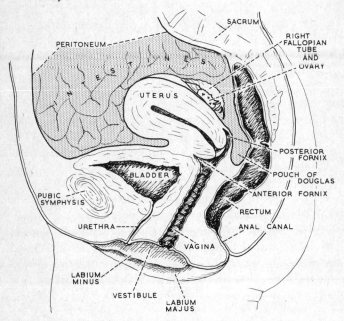

Fig. 19/2 A section through the female pelvis

2. *The formation of female sex hormones* The female sex hormones are produced in the ovarian follicle and its successor, the corpus luteum (see p. 192).

The *uterine (Fallopian) tubes*, right and left, are thin tubes attached at their inner ends to the uterus, into which they open (see Fig. 19/3). They lie in the upper border of the broad ligament of peritoneum at the sides of the uterus. At its outer end each tube has a small opening into the peritoneal cavity. This opening is surrounded by long moving processes called fimbriae. At ovulation these fimbriae surround the spot on the surface of the ovary where the ovarian follicle is breaking through and cause the ovum to enter the tube. The interior of the tube is lined with ciliated cells, and by the movements of these cilia the ovum is conveyed along it and into the uterus, a journey that takes several days. If sexual intercourse takes place during this time, the spermatozoa reach the ovum as it passes along the tube and one of them may fertilise it. If it is not fertilised, it passes into the uterus and there dies.

The *uterus* (womb) is a hollow, thick-walled, muscular organ about 7.5cm long. It lies in the pelvic cavity with the bladder below and in front of it, the rectum behind it, and coils of intestine above it (see Fig. 19/2). The body of the uterus is the broad upper two-thirds, the cervix is the narrower and cylindrical lower one-third. The fundus is the rounded upper end of the body. The uterine tubes open into the upper end of the body, just below the fundus. The cervix projects backwards into the upper end of the vagina, into which it opens, the opening being called the external os. The front and back of the uterus are covered with peritoneum, which passes off its sides as the broad ligaments, between whose double layers run the uterine tubes and the blood-vessels, lymph-vessels and nerves of the uterus.

The uterus is composed of three coats: (a) an outer layer

of peritoneum; (b) a thick muscular coat of many interlaced bundles of unstriped muscle fibres, enormously increased in number and size during pregnancy, and (c) the endometrium, the inner mucous membrane, which undergoes monthly cyclical changes. The cavity of the uterus is lined by this endometrium; in childhood and when a woman is not pregnant, the cavity is little more than a chink with its anterior and posterior walls almost touching.

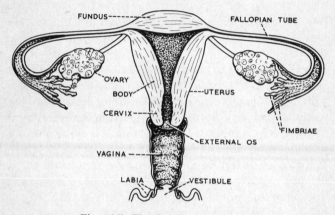

Fig. 19/3 The female genital organs

The *vagina* is a muco-muscular tube about 7.5cm long in front and about 9cm long behind. It is shorter in front because of the projection of the cervix into its anterior wall (see Fig. 19/2). The posterior fornix is the part of the vagina above the cervix, the anterior fornix is the part below it. In front of the vagina is the urethra, behind it are the rectum and anal canal. Its lower end opens on to the surface of the body at the vestibule between the two labia minora.

The *external genitalia* are called the vulva. Two *labium majorum,* long prominent folds of skin, cover two *labium minorum,* smaller folds surrounding the vestibule, into which the urethra opens in front and the vagina behind. The hymen, a thin fold of mucous membrane, may partly close the lower end of the vagina.

THE FEMALE SEXUAL CYCLE

From puberty at the age of about 11 or 12 years up to menopause (change of life), one ovum reaches maturity every month in one or other ovary. In the uterine tube this ovum may be fertilised by the penetration of a spermatozoon and a pregnancy is started, or it is not fertilised and dies.

After the ovum has been discharged, the ovarian follicle is converted into a *corpus luteum*, a rounded mass of special cells, which for eight days increases in size, its cells producing *progesterone*, a hormone. If the ovum is not fertilised, the corpus luteum starts to degenerate, its cells shrink, and in a few weeks it becomes a scar on the surface of the ovary. If the ovum is fertilised, the corpus luteum continues to develop and to produce more progesterone. It reaches its maximum size about the fourth or fifth month of pregnancy, being then about 5.0cm in diameter, and during the rest of the pregnancy it gradually becomes smaller. The progesterone it produces prepares the endometrium to receive a fertilised ovum and stimulates the growth of breast tissue.

Between puberty and the menopause the ovaries show all stages of activity: the gradual development of ova within follicles, the monthly ovulation, the degeneration or growth of the corpus luteum, and the scars at the sites of old corpora lutea.

Changes in the Endometrium
During the reproductive period the endometrium of the uterus undergoes cyclical changes under the influence of

oestrogens produced by the ovarian follicle and progesterone produced by the corpus luteum, the production of these hormones being controlled by hormones produced in the hypothalamus-pituitary complex. The purposes of the endometrial changes are to prepare the endometrium to receive a fertilised ovum or to cause it to be shed if the ovum is not fertilised.

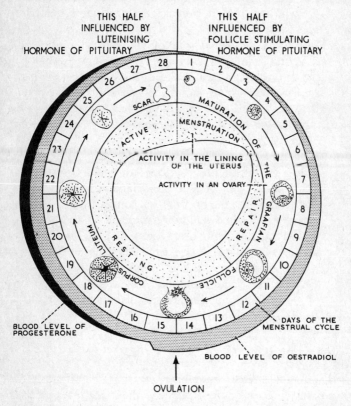

Fig. 19/4 A diagram of the menstrual cycle and female hormonal activity

The average menstrual cycle takes 28 days, the regularity being due to the regular occurrence of ovulation (see Fig. 19/4). The changes that take place in the endometrium are divisible into three phases:

1. *Proliferative or pre-ovular phase* which lasts for about 14 days and is controlled by oestrogens. It lasts from the end of menstruation to the discharge of the ovum from the ovary. There is a rapid regrowth of endometrium, the whole of the interior of the uterus being relined in two days.

2. *Secretory or post-ovular phase* which lasts for about 13 days and is controlled by progesterone and a little oestrogen produced by the corpus luteum. The endometrium becomes thick and more vascular, and its glands become dilated with secretion. If the ovum is not fertilised, the amount of progesterone starts falling about the 22nd day.

3. *Menstrual phase* which lasts for about four to five days. The secretion of progesterone continues to diminish. The endometrium degenerates, glandular secretions are discharged, and the endometrial capillaries break down and bleed. *Menstrual fluid* is composed of blood, degenerated endometrium, and secretions from endometrial and cervical glands. It does not clot because of the presence of endometrial tissue.

Ovulation occurs about the 14th day after the first day of menstruation, and sometimes causes a little pain.

THE BREASTS

The female breasts (mammary glands) secrete milk after childbirth. Small before puberty, they then enlarge to a variable size; they enlarge further during pregnancy and lactation, and atrophy in old age. The nipple is encircled by the areola, a circle of pigmented skin, and traversed by the ducts of the breast.

The breasts are composed of lobes of glandular tissue, separated by partitions of fibrous tissue and covered with fat (see Fig. 19/5). Each lobe is composed of the cells which during lactation secrete milk. The milk passes along ducts, which converge radially towards the nipple on which they open.

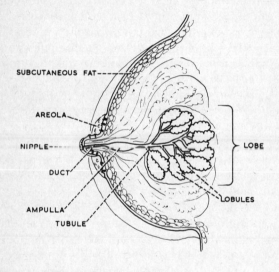

Fig. 19/5 The structure of the breast

In pregnancy the development of the breasts is due to high levels of progesterone and oestrogen in the blood. After birth colostrum, a thin yellow fluid containing fat and protein is secreted for a day or two and is followed by milk, the breasts being stimulated to produce it by prolactin, a hormone produced by the pituitary gland.

In males the breasts are rudimentary and functionless.

HEREDITY

Heredity is the transmission of physical and mental characteristics from parents to their children.

A person is the result of two factors, heredity and environment. The hereditary factor is determined for him at his conception, the fertilisation of his mother's ovum by his father's spermatozoon.

To understand how inherited factors are acquired from each parent, we must study cell division. *Mitosis* is the ordinary division of one cell into two cells, a process performed by the splitting of the nucleus and then the splitting of the cell. *Chromosomes* are organised structures present in the nucleus of every cell. When a cell is not in process of dividing, they appear as a tangle of thin threads, but when the cell is about to divide they appear as distinct rod-like structures. In human cells there are 46 chromosomes. Just before the nucleus divides, each chromosome divides lengthwise into two equal parts, of which one part goes into the nucleus of each new cell, which accordingly has the right number of chromosomes.

Meiosis is the special form of division that occurs only in the formation of ova and spermatozoa. By it, each ovum and spermatozoon gets only half the usual number of chromosomes, with the result that to the fertilisation of an ovum the ovum itself brings 23 chromosomes and the spermatozoon 23, the fertilised ovum then having the right number, 46, derived equally from both parents.

Genes are the factors of inheritance and form the chromosomes. For each inherited characteristic there are a pair or several pairs of genes. Each gene is thought to be responsible for one particular type of enzyme reaction in a cell. Each gene in a pair may have similar tendencies, e.g. to produce right-handedness, or each may have different tendencies, e.g. one to produce right-handedness and one to produce left-handedness. Genes are inherited according to definite rules. A *dominant gene* is one which is able to suppress the appearance of the characteristics of the other, which is called a *recessive gene*.

If the chromosomes of either parent contain a dominant gene, it is transmitted to half his or her children. This half shows the particular feature produced by that gene and will transmit it to half their children. The other half will not inherit the dominant gene and cannot suffer its effects nor transmit it to their children. The degree to which a dominant gene will produce an effect is variable. The 'expression' of a gene (i.e. its ascertainable effects) may be full or less than full, and occasionally a gene appears to 'skip' a generation and produce its effects in the next.

A recessive gene must be present in both parents if it is to produce effects. Any child of such parents has a one in four chance of inheriting the condition it produces. If only one parent carries a recessive gene, any child of his or hers will inherit it without showing any effects and will transmit it to his or her children. The chance of inheriting the same recessive gene from both parents is much increased if they are cousins.

Sex Inheritance

The sex chromosomes are the special chromosomes that decide sex. In *females* there are two X chromosomes in the cells. In *males* there is a small Y chromosome and an X chromosome, identical with the X chromosome in a female. By the process of meiosis, an *ovum* has 22 ordinary chromosomes and one X chromosome; a *spermatozoon* has 22 ordinary chromosomes and *either* a Y chromosome *or* an X chromosome.

The sex of a child depends upon whether the sperm which does the fertilising carries a Y or an X chromosome. For in the fertilisation of an ovum *either*, an ovum with an X chromosome is fertilised by a spermatozoon with a Y chromosome, and a male child is produced because the sex chromosomes are XY; *or*, an ovum with an X chromosome is fertilised by a spermatozoon with an X chromosome, and a female child is produced because the sex chromosomes are XX.

Chapter 20

DEFENCE MECHANISMS

Throughout life a person is exposed to many dangers – in particular to injuries, infections and new growths. Against injury and infections he has been able to develop defence mechanisms, but he has not been able to develop adequate natural defences against new growths.

Against *injuries* he has a set of defence mechanisms, including automatic, rapidly acting reflex mechanisms. Painful stimulation of any part will produce an immediate muscular response by which the affected part is jerked away from pain and danger. The eye protects itself by blinking or shutting when anything approaches it too quickly and by producing tears when anything lands on the conjunctiva. Sneezing and coughing expel dust and other particles from the nose, pharynx and larynx. Poisonous substances entering the stomach are often ejected by vomiting.

Against *infections* man is protected by an elaborate system of defences. Throughout life he is exposed to harmful micro-organisms; he breathes them in, they settle on his skin; they are present in many of the things he eats and drinks; and to survive their attacks he must have adequate defences.

Against micro-organisms settling on the skin he has a defence in the skin itself, through which micro-organisms cannot easily pass unless the skin is scratched, cut or punctured. The mucous membranes, such as those of the lips, are more easily penetrated. The eyes are protected by the tears, which can wash micro-organisms away and destroy some with lysozyme, an anti-bacterial substance present in tears.

Against the micro-organisms we breathe in, we have three methods of defence:

1. The sneezing and coughing with which we try to expel from the air-passages anything that should not be there.
2. The mucus fluid secreted by the cells of the mucous membrane lining the air-passages, a fluid through which micro-organisms cannot easily pass.
3. The cilia of the cells on the surface of the mucous membrane which flick away from the lungs any particles that land on them.

Against micro-organisms in food and drink we have the defences of the saliva, which contains lysozyme, and the acid gastric juice. Between them they deal so effectively with micro-organisms that few get into the intestine, but among some that get in are some dangerous ones, such as those that cause typhoid fever and cholera.

Inflammation

Inflammation is the reaction of tissues to infection. A person who gets a boil on his skin is aware that the affected part is hot, painful, red and swollen. The pain is due to swelling of the tissues caused by a discharge of fluid into the tissue spaces. Pain is a useful and at times vital function of tissues in danger, a warning that something is wrong. The heat and redness of the boil are due to a dilatation of the arteries and capillaries in the area, which ensures that more blood is provided. The fluid in the tissue spaces is rich in antibacterial and antitoxic substances, capable of destroying the micro-organisms and of neutralising the toxins (poisons) some of them produce.

The *granulocytes* of the blood respond to infection by increasing in numbers and being attracted towards the affected part, where they pass through the capillary wall into the tissue spaces, stimulated to do so by chemical substances in the damaged tissues and in the micrc organisms. During an acute infection the number of them can increase from a normal 2 500–7 500 per mm³ of blood to 50 000 per mm³ or more. The granulocytes attack the micro-organisms, absorb them into the cell and there destroy them. An *abscess* may form in the centre of the boil. It contains *pus* – a creamy or green fluid containing dead and living cells and micro-organisms. With healing of the boil, any dead micro-organisms and cells are absorbed by the scavenging cells of the reticulo-endothelial system, and the part returns to normal.

THE DEVELOPMENT OF IMMUNITY

Immunity is the body's ability to withstand infection by harmful micro-organisms. Immunity can be:

1. A racial immunity acquired by members of a race as a result of attacks of the infection in their ancestors.
2. A temporary immunity transferred from mother to child, both before birth through the placenta and in the first few days of breast feeding.
3. An immunity acquired as a result of a previous infection by the same micro-organism.
4. An immunity acquired artificially by giving a person a vaccine or a toxoid. A vaccine is a preparation of a micro-organism injected into a person or taken by mouth. A toxoid is a modification of a toxin produced by a micro-organism; it is harmless but capable of producing an immunity response.

The duration of an acquired immunity is variable. An attack of some infections (e.g. mumps, measles) can produce life-long immunity; an attack of others (e.g. common cold, pneumonia) produces only a short immunity. The influenza virus has the ability to change its chemical composition, with the result that an immunity to one strain of it may not provide protection against another. An immunity provided by vaccination can last a variable length of time and is not usually as effective as an attack of the infection itself.

Immunity is due to the action of an *antigen* in stimulating the body to produce an *antibody*.

It is necessary to understand first the actions of the thymus and the lymphocytes.

The Thymus

The thymus is a soft, multi-lobed gland, which lies in the front of the neck and upper part of the thorax. It is well developed at birth, increases in size up to about sixteen years, and then becomes smaller. It is composed of lymphocytes, clumps of special cells called thymic corpuscles, and connective tissue.

It is an essential organ in the development of cell-mediated immunity (see below) and for its maintenance in later life, and it is responsible for the 'processing' of many lymphocytes.

Lymphocytes

Lymphocytes are one form of white blood cell. Normally there are 1 500–2 700 per mm³ of blood. They are small, round cells with a large round nucleus which occupies most of the cell. They are present in large numbers in lymph-nodes and other lymph-tissue, such as the tonsil. Some are produced in bone marrow and live for only a few days. Others which live for months and possibly for years are produced in the thymus and other lymph-tissue after developing from cells in bone marrow.

There are two kinds of these long-lived lymphocytes.

T lymphocytes go out into the blood and tissue-spaces to

attack invading micro-organisms. They have to go into the thymus to be 'processed', i.e. treated in such a way that they are able to take part in immunological reactions. They play a role in cell-mediated immunity and stimulate the B lymphocytes to produce antibody.

B lymphocytes turn themselves into another kind of cell called a plasma cell and remain mostly inside lymph-nodes and lymph-tissue. When an antigen comes into contact with a B lymphocyte, the lymphocyte produces antibody, but to do this for some micro-organisms it has first to be stimulated by a T lymphocyte.

In combating an infection there is a great deal of co-operation between the two kinds of long-lived lymphocytes.

Antigen

An antigen is a chemical substance that causes an immunological response when it gets into the body or comes into contact with the skin. An antigen produces this reaction because it is 'foreign' and is recognised by the cells of the body to be foreign. Harmful micro-organisms contain antigens. When a micro-organism attacks the body, the body has two ways of rendering the antigen harmless or destroying it, and of developing immunity to any subsequent attack by the same organism. The ways are: antibody formation and the development of cell-mediated immunity.

1. *Antibody formation*

Antibodies are immunoglobulins (a form of protein) produced by the presence of an antigen. When an antigen arrives in lymph-tissue, the plasma cells (which have developed from B lymphocytes) produce an antibody to that antigen. The antibody is a specific one, e.g. the antibody produced in response to the micro-organism that causes typhoid fever is different from the antibodies produced in response to infection by other kinds of micro-organism. The antibodies circulate in the blood and are a principal defence against that particular micro-organism.

Some micro-organisms (such as the one that causes diphtheria) produce their harmful effects by releasing from their interior, a toxin. The toxin passes into the blood, and to neutralise its effects a specific antibody called an antitoxin is produced.

2. *Cell-mediated immunity*

This kind of defence is carried out by the lymphocytes themselves, which travel to meet the antigen wherever it is and there destroy it. This is an important function of lymphocytes when the antigens do not readily enter lymphtissue.

Harmful Effects of Immunological Reactions

Not all immunological reactions are beneficial. Some can be harmful.

Immunity is involved in the ability a body possesses to recognise its own 'self' cells and the 'non-self' cells of someone else or of another type of animal. It is not normally possible (except in identical twins, who are immunologically identical) to graft successfully a tissue from one person to another. One's own tissue can be grafted on to one's own body – e.g. skin from the arm can be used to repair damage to the face; but grafting skin or a kidney or heart or other organ on to anyone else (except an identical twin) is doomed to failure unless the recipient has been prepared to receive it by special techniques aimed at preventing the antigen-antibody reaction from taking place. They do not always succeed, for the cells of one's own body can recognise foreign cells and set out at once to destroy them.

Immunological reactions are also involved in allergic conditions (such as asthma, hay fever and skin reactions to substances to which a person is 'sensitive'), in blood transfusion accidents when blood of the wrong type is given, and in some diseases called auto-immune diseases in which the body appears to be unable to recognise some of its own cells and treats them as if they were foreign cells from another body.

INDEX

Abdominal arteries 80–1
  cavity 52–3
  wall 52
abducens nerve 153
absorption of food 107–8
accessory nerve 154
accommodation, in eye 172
acetabulum 49, 65–6
acetylcholine 36
adenoids 144
adipose tissue 16
adolescence 30
adrenal cortex 182
  gland 181–3
  medulla 182–3
adrenaline 182–3
adrenocorticotrophic hormone
    (ACTH) 178
adult life 30
air, composition of 90–2
aldosterone 182
alimentary tract 100–12
  development of 22
  hormones produced by 185
allergic conditions 204–5
alpha rhythm 161
alveolus 89
amino-acids 93–4, 113
ampulla of Vater 123
anal canal 109
  sphincter 109, 110
androgens 182, 184

ankle joint 68
anterior tibial artery 70–1
antibodies 200, 203–4
antidiuretic hormone 180
antigen 200, 203–4
anus 109
aorta 80
aortic valve 77
appendix 109
aqueous humour 170
arachnoid membrane 158
arches of foot 28, 68–9
areolar tissue 16
arm 13, 57–64
  development 22
arteries 79–81
arterioles 81
arytenoid cartilages 87
ascorbic acid 116
association areas, of brain
    150–1
atlas vertebra 45
atrio-ventricular bundle 77
atrium, left 76–7
  right 76
auditory meatus, external 174
  nerve 154
  ossicles 174–5
auricle 174
autonomic nervous system
    145–6, 164–7
axillary artery 63

axis vertebra 45
axon 147

Back, muscles of 55
balance 175
basal ganglia 151
base of skull 39
basilar artery 159
bile 124–5
  duct 123
biliary system 122–3
birth 24
bladder, urinary 130
blind spot 169, 173
blood 92–8
  cells 94–6
  circulation of 73–84
  clotting 97
  groups 97–8
  plasma 93–4
    proteins 93, 124
  platelets 96
  pressure 81, 131
  vessels 79–84
bolus 102
bone 16, 31–3
  cancellous 31
  compact 31
  development 26–7
  marrow 32–3
Bowman's capsule 128–9, 132
brachial artery 63
  plexus 64, 156
brachiocephalic artery 80
brain 147–53
  blood supply 159–61
  development 26
  stem 153
breasts 194–5
broad ligament 112
bronchiole 89

bronchus 87
bundle of His 77

Caecum 108
calcaneum 65
calcium 116–7
calories 118
capillaries 82
carbohydrates 114–5, 123–4
carbon dioxide 90–2
cardiac cycle 77–9
  muscle 17, 75
carotid arteries 42, 159
  bodies 90
carpus 59
cartilage 16
  costal 48
  laryngeal 86–7
  semilunar, of knee 68
cauda equina 154
cells 13–4, 18–9
cellulose 114
cerebellum 148, 152, 153
cerebral arteries 159–60
  peduncles 151–2
  hemispheres 148–51
cerebrospinal fluid (CSF) 156–8
cerebrum 148–51
cervical vertebrae 45
cervix uteri 190
chest 47–9
cholecystokinin 124–5, 185
chorionic gonadotrophin 184
choroid coat 169
chromosomes 196–8
chyme 107
ciliary body 169
circle of Willis 159
circulation of blood 73–84
clavicle 57–8
clotting of blood 97

cobalt 95, 117
coccyx 46
cochlea 175
colon 109
colostrum 195
cones, of retina 170, 172–3
conjunctiva 170–1
copper 95, 117
cornea 168
coronary arteries 79
corpus callosum 149
    luteum 184, 192
    striatum 151
cortex, cerebral 148, 151
cranial nerves 153–4
    vault 39–40
creatinine 131
cricoid cartilage 87
cystic duct 122, 123
cytoplasm 14

Dark adaptation 173
defaecation 110
defence mechanisms 199–205
dentition 28–9
deoxyribonucleic acid (DNA)
    14
dermis 136–7
diaphragm 48–9, 90
diastole 78
diet, balanced 118–9
digestion 104–5, 107–8
disaccharides 114–5
ductus arteriosus 25
duodenum 105
dura mater 159

Ear 174–5
ectoderm 21
ejaculatory duct 187–8
elbow joint 60–1

electrocardiogram   (ECG)
    78–9
electroencephalogram (EEG)
    161
embryo 21–4
endocardium 73
endocrine glands 177–85
endoderm 22
endometrium 191, 192–4
enzymes 14
epidermis 136
epididymis 186–7
epiglottis 87
epiphysis 27
epithelium 15–6
    germinal 188
erector spinae muscles 55
erythrocyte 94
erythropoietin 185
expiration 90
external   female   genitalia
    191–2
eye 168–73
eyelids 170

Facial nerve 153
faeces 110
fats 94, 113–4, 124
female genital organs 188–92
    sex hormones 190
    sexual cycle 192–4
femoral artery 70
    vein 71
femur 64
fertilisation of ovum 20–1
fetus 24
fibrocartilage 16
fibrous tissue 16
fibula 65
folic acid 95, 116
follicle-stimulating   hormone
    (FSH) 178

fontanelles, of skull 28
foot, arches of 28, 68–9
  bones of 65
foramen ovale 24
forebrain 148–51
frontal lobe 149

Gall-bladder 122–3
gastric juice 104–5
  movements 105
gastrin 185
genes 196–7
germinal epithelium 188
glomerulus 128–9, 132
glossopharyngeal nerve 154, 176
glucagon 126, 183
glucocorticoids 182
glycogen 115, 123–4
gonadotrophic hormones 178–9
granulocytes 95–6, 202
great omentum 112
growth, after birth 25–30
  before birth 20–4
  hormone (GH) 178

Haemoglobin 92, 94
hairs 137–8
head 38–43
hearing 150, 173–6
heart 73–9
  development 22
  sounds 78
heat gain and loss 140
hemispheres, cerebral 148–151
hepatic ducts 122
heredity 196–8
hind-brain 152–3
hip bone 49–51
  joint 65–6

histiocytes 14
hormones 94, 177–85
humerus 58
hydrocortisone 182
hyoid bone 40–1
hypoglossal nerve 154
hymen 192
hypothalamus 151, 176
hypothalamus-pituitary complex 176–80

Ileum 106–7
ilium 49
immunity, cell-mediated 204
  development 202–4
incus 174
inferior vena cava 83
inflammation 202
inguinal ring 187
inheritance, of sex 197–8
innominate bone 49–51
inspiration 90
insulin 126, 183
internal carotid artery 42, 159
  jugular vein 43
interstitial-cell stimulating hormone (ICSH) 179, 184
intervertebral discs 46
intestinal juice 107
intestine, large 108–10
  small 106–8
involuntary muscle 17
iodine 117
iris 168–9, 171, 172
iron 117
ischium 49
islets of Langerhans 126

Jejunum 106–7
joints 33–5
jugular veins 43

Kidneys 127–9, 131–2, 185
  development 23
knee-joint 67–8

Labia majora 191
  minora 191–2
lacrimal gland 171
lacteals 142
larynx 86–7
leg 13, 28, 64–72
  development 22
lens, of eye 170, 172
levator ani 51
ligamentum teres 66
liver 121–5
  development 22–3
loop of Henle 129, 132
lumbar vertebrae 46
lungs 87–92
  development 22
luteinising hormone (LH) 178–9, 184
lymph 144
  nodes 141–2
  tissue 143–4
  vessels 141
lymphatic system 141–4
lymphocytes 96, 201, 203, 204

Magnesium 117
male genital system 186–8
malleus 174
mammary glands 194–5
mandible 40
maxilla 40
medulla oblongata 152–3
medullary cavity, of bone 31
meiosis 196
Meissner's corpuscles 138
melanocyte stimulating hormone 180
melatonin 185

meninges 158–9
menopause 30
menstrual cycle 194
  fluid 194
mesentery 111
mesoderm 21–2
metabolism 113–20
metacarpals 59
metatarsals 65
micturition 133
midbrain 151–2
middle ear 174–5, 176
milk 119–20, 195
mineralo-corticoids 182
minerals, of body 116–17
mitosis 196
mitral valve 77
monocytes 96
monosaccharides 114–5
motor area of brain 149
  end plate 36
  impulses 145, 162–3
  nerves 145, 162–3
mouth 100–1
muscle 16–7, 35–7
myelination 26
myocardium 75
myofibrils 35

Nails 137
nasal bones 39–40
  cavities 85–6
naso-lacrimal duct 171
nephron 127–9
nerve-fibres 146–7
nervous system 145–67
  development 22
neurone 18, 146–7
  lower motor 163
  upper motor 163
neuroglia 18, 146
neuromotor unit 36

nicotinic acid 116
nor-adrenaline 183
nose 85–6
nucleolus 14
nucleus, cell 14
nutrition 113–20

Oculomotor nerve 153
oesophagus 102
oestrogens 184
old age, changes in 30
olfactory nerve 153
optic disc 169
    nerve 153, 170
organ of Corti 175, 176
osmotic pressure 134–5
ossicles, auditory 174–5, 176
ossification 26–7
ovarian follicle 184, 189, 190
ovary 184, 188–90
    development 23
ovulation 189, 194
ovum 188–9, 190, 192
    fertilisation 20–1, 190
oxygen 90–2
oxyhaemoglobin 92, 94
oxytocin 180

Paccinian corpuscles 138, 139
pancreas 125–6, 183
    development 22–3
pancreatic juice 125–6
parasympathetic nervous system 165, 166–7
parathyroid glands 181
    hormone 181
parietal lobe 149
parotid gland 101
patella 64
pelvis 28, 49–51
penis 188

pericardium 75
periosteum 31–2
peritoneal cavity 110
peritoneum 110–12
perspiration, insensible 140
Peyer's patches 144
phalanges, of fingers 59
    of toes 65
pharyngeal-tympanic (Eustachian) tube 174
pharynx 86, 100
phosphorus 117
pia mater 158
pineal gland 185
pituitary gland 177–80
placenta 21, 184
plasma 93–4
    proteins 93, 124
platelets 96
pleura 88–9
polymorphs 95–6
polysaccharides 114
pons 152
popliteal artery 70–1
portal veins 84
posterior tibial artery 71
posture 55
potassium 117
prefrontal lobes 150–1
premotor area 149
progesterone 184, 192
prolactin 180, 195
prostate gland 130, 188
proteins 113, 124
    plasma 93, 124
puberty 29–30
pubic bones 49
pulmonary artery 76
    valve 76
    veins 76–7
pulse 81
pupil, of eye 168, 172

pus 202
pyloric canal 103
pyramidal decussation 153

Radial artery 63
radius 59
receptaculum chyli 142
reciprocal innervation 36
recto-uterine pouch (of Doug-
    las) 112
rectum 109
red blood cells 94–5, 124
reflex actions 163–4
renal artery 127
reproduction 186–98
respiration 85–92
respiratory centre 90, 92
reticular system 153
reticulo-endothelial   system
    99
retina 169–70, 172
Rhesus factor 98
riboflavin 116
ribonucleic acid (RNA) 14
rods, of retina 170, 172
Rolando, fissure of 149

Sac, peritoneal 111
sacrum 46
saliva 101
salivary glands 101
salts, in plasma 94
scapula 58
sciatic nerve 72
sclera, of eye 168
scrotum 186
sebaceous glands 137–8
sebum 138
secretin 185
semicircular canals 175
seminal vesicle 187
seminiferous tubules 186

sensation 149–50
sensory nerves 145, 161
    spots 138–9
serum 97
sex gland hormones 178–9
shoulder joint 59–60
sight 150
sinu-atrial node 77
sinusoids 82
skeletal tissue 16
skeleton 33
skin 136–9
skull 38–40
    development 28
small intestine 105–8
smell 150
    organ of 176
sodium 182
soft palate 102
sound 175–6
special senses 168–76
speech 150
    centre 150
spermatic cord 187
spermatozoon 20, 186
spinal column 43–7
    cord 154
    ganglions 155
    nerves 155–6
spiral organ 175, 176
spleen 98–9, 144
stapes 174–5
sternum 48
stomach 102–5
subclavian artery 62–3
sublingual gland 101
submandibular gland 101
superior vena cava 83
swallowing 101–2
sweat 140
    glands 138
Sylvius, fissure of 149

sympathetic nervous system 165, 166–7
synapse 147
synovial fluid 34
  membrane 34
systole 78

Talus 65
tarsal plate 170
tarsus 65
taste 150
  buds 100, 176
  organ 176
tears 171
teeth 41
  development 28–9
temperature of body 139–40
temporal bone 40
  lobe 149
tendon 35
testis 184, 186
  development 23–4
testosterone 184
thalamus 151
thiamine 116
thoracic duct 142–3
  vertebrae 45
thorax 47–9
thoroughfare vessel 82
thymus 144, 201
thyro-calcitonin 180, 181
thyroid cartilage 86–7
  gland 180
thyro-trophic stimulating hormone (TSH) 178, 180
thyroxine 178, 180
tibia 64–5
tibial arteries 70–1
tongue 100–1
tonsils 143
touch 138–9
trachea 87

tricuspid valve 76
trigeminal nerve 153, 176
tri-iodothyronine 178, 180
trochlear nerve 153
tunica vaginalis 186
twins 21
tympanic membrane 174, 175–6

Ulna 59
ulnar artery 63
  nerve 58–9
urea 94, 124, 131
ureters 129–30
urethra, female 131
  male 130–1, 188
uric acid 124, 131
urinary bladder 130
  system 127–33
urine 131–2
uterine (Fallopian) tube 190
uterus 190–1

Vagina 191
vagus nerve 154
vas deferens 187
veins 82–4
vena cava, inferior 83
  superior 83
venous sinuses, of skull 159
ventricles, of brain 156
  of heart 76, 77
venules 82
vertebrae 44–7
vertebral artery 159
  column 43–7
vision 168–73
visual purple 173
vitamins 115–6
  A 115, 124, 173
  B 116
  B_{12} 95, 116

C 116, 182
D 115, 124, 131
K 115, 124
vitreous humour 170
vocal cords 87
voluntary muscle 17, 35–7
vulva 191–2

Walking 70
water 117–8
  balance 133–4

weight, at birth 26
white blood cells 95–6
wisdom teeth 29
wrist, bones 59
  joint 61

Y-shaped ligament 66

Zygomatic bone 40